WHEN THE SMOKE CLEARS
STORIES AND REFLECTIONS OF A WISCONSIN FOREST RANGER

BLAIR W. ANDERSON

LITTLE CREEK PRESS
MINERAL POINT, WISCONSIN

Copyright © 2024 Blair W. Anderson

All rights reserved. No part of this publication may be reproduced, distributed, or transmitted in any form or by any means, including photocopying, recording, digital scanning, or other electronic or mechanical methods, without the prior written permission of the publisher, except in the case of brief quotations embodied in critical reviews and certain other noncommercial uses permitted by copyright law. For permission requests or other information, please send correspondence to the following address:

Little Creek Press
5341 Sunny Ridge Road
Mineral Point, WI 53565

ORDERING INFORMATION
Quantity sales. Special discounts are available on quantity purchases by corporations, associations, and others. For details, contact info@littlecreekpress.com

Orders by US trade bookstores and wholesalers.
Please contact Little Creek Press or Ingram for details.

To contact the author, Blair Anderson: whenthesmokeclears2024@gmail.com

Printed in the United States of America

Cataloging-in-Publication Data
Name: Blair W. Anderson, author
Title: When the Smoke Clears. Stories and Reflections of a Wisconsin Forest Ranger
Description: Mineral Point, WI Little Creek Press, 2024
Identifiers: LCCN: 2024918416 | ISBN: 978-1-955656-82-5
Classification: Biography & Autobiography / Fire & Emergency Services
Biography & Autobiography / Law Enforcement
Biography & Autobiography / Memoirs

Book design by Little Creek Press

Cover photo: Chris Klahn/Mike Lehman

To my parents, now long departed,
for their example, pointing me toward the
One who makes all things possible.

PRAISE FOR
WHEN THE SMOKE CLEARS

"*When the Smoke Clears* gives the reader a taste of what wildland firefighters and foresters experience in their important work protecting lives and property and conserving our vital natural resources. Blair recounts a diverse and eclectic array of stories, some sobering and others humorous, emphasizing the many lessons he learned from a career in public service. Blair also reveals the importance of community and the power of meaningful relationships."

—Paul DeLong, former Wisconsin chief state forester

"Blair Anderson was employed by the Department of Natural Resources for many years during the time I represented much of southwestern Wisconsin in the state legislature. Our paths frequently crossed because I was a volunteer firefighter for ten years and a member of Natural Resources Committees throughout my career. Blair's life might have turned out very differently had he not chosen to follow a hunch and radically change his career path. For the curious among us, Blair's book offers us a chance to glimpse vicariously the proverbial road not taken in our lives and muse how our lives might have turned out differently if we had followed our own hunches."

—Dale Schultz, former Wisconsin state senator, 17th District

"Blair writes with new insight into the calling of a forest ranger in Wisconsin. This job is not about smoke and flame or red lights and sirens, nor is it about the shadowy world of the arsonist or the calculating timber thief. Blair writes about the real workings of the world of the ranger. The truth is that this calling is really about people's lives and the people themselves—the good, the bad, and the ornery. With his great humanity, Blair captures the inside job."

—Ed Forrester, former Wisconsin Department of Natural Resources area forestry supervisor, Cumberland

"This book provides the combined insight into the life and times of a DNR forester with the flavor of an autobiography. The contents take the reader through the complexities involved behind the scenes with challenging DNR activities that most people never realize or appreciate. Blair provides readers with the human side of DNR life along with the logistical and strategic decisions related to the wide range of work tied to the protection of Wisconsin's natural resources. Additionally, Blair shares his personal journey that led up to his career with the DNR.

Blended into this narrative of short stories are Blair's personal life decisions that created the unusual life path that took him from the world of big-city business employment to falling in love with rural American life and work in southern Wisconsin. The book takes the reader into the mind of one man's life journey in following his passion for hard work, simplicity, and understanding what it means to never "work" a day in your life when you enjoy your job and love your family.

Anderson writes his short stories with humor and captures the reader's interest by including history and nostalgia. Topics range from forest fires and floods to cantankerous landowners and freeing a bald eagle from a ground trap. Embedded within the book are references to the many good people who represent strong community values, Christian leadership, and those who help a neighbor in need."

—Jamie Benson, former River Valley School District superintendent

"Now retired, Blair Anderson takes a positive look back on the decision he made in his youth to redirect his career path. While foresters are often seen as committed to a solitary existence in the natural world, Anderson reminds us that the "fire job" in forestry demands people who can engage the human ecosystem in community service. It demands people who can prepare response capability to fit the measured need for a "game day" selected by the weatherman. These stories record the fact that a high percentage of workdays hold the promise of work that is fun—with serious potential for an adrenaline buzz—in the company of friends."

—John Grosman, retired USFS regional wildland fire training officer

"Based on my twenty years of experience as a volunteer fire chief/EMT, I strongly endorse Blair's description of his career. Furthermore, as our department is adjacent to the Spring Green Ranger Station, I greatly appreciate Blair's accurate description of the area and individuals that I'm familiar with. Having known Blair for a long time in various settings, I can say with confidence that he's the real deal. A very enjoyable read. Well done!"

—Eric Drachenberg, retired Arena Fire Department chief/EMT

"Blair Anderson's *When the Smoke Clears* offers vivid storytelling, wry humor, and humble insights into a world so many know only as a distant and romantic notion. Readers will not only learn what it takes to protect the great American outdoors from raging forest fires but also lessons about self, the mythologies we carry, and a deepening rural-urban divide that will only close if we believe in things as much as Anderson believed in his work."

—Brian Reisinger, author of *Land Rich, Cash Poor: My Family's Hope and the Untold History of the Disappearing American Farmer*

"I found *When the Smoke Clears* both interesting and informative. Since Blair and I served together for a period of time, it was interesting to see how the Wisconsin DNR differed from the Indiana DNR. There were actually more similarities than differences. It was hard to separate my role as "reviewer" from my self-imposed role as "reader." While I found the fire stories interesting, my background led me to the conclusion well before I read all the words. When you have intimate knowledge of a subject matter, you tend to be a bit pickier.

While some of the "fire stuff" brought back fond memories or conjured up stories from my own past, it was the anecdotal stories that kept me glued to the manuscript. In most cases, your stories brought back a flood of memories from my own career. Many of those were quite similar and often with the same conclusion."

—Steve Creech, retired Indiana state fire supervisor

TABLE OF CONTENTS

PREFACE 1

SETTING THE STAGE 3
   IN THE BEGINNING 4

BECOMING A FOREST RANGER 12
   THE MYSTIQUE 13
   REALITY CHECK 13
   IN TRAINING 15
   WE'RE MOVING WHERE? 16
   THE SPRING GREEN JOB 17
   WHO KNEW? 19
   MAYBE THIS FITS OKAY 19
   YOU WANT ME TO HELP WITH WHAT? 21
      Dedicated 21
      An Injured Critter 22
   HELPING AND FREEDOM 25
      The Dotted Line 27
   DEVIATIONS: SOMETHING'S CHANGED 30

A FOREST FIRE PRIMER 33
   THE BASICS 34
   PUTTIN' OUT THE FIRE 36
   WHY WAS THAT SO TOUGH TO STOP? 36
   HOW IT WORKS IN WISCONSIN 39

PREPARING FOR WINTER 41
TIME TO GO TO WORK 42
    Louie. 45
    Up In Smoke 46
IT'S AS SIMPLE AS THAT 47
SOOOO BIG 48
    Squeaky. 49
    Take Cover! 52
    Night Fires in a Marsh 56
THE MEASURE OF A RANGER 60
    Marble Quarry Road 61
    How Does Your Garden Burn? 64
FIRE DEPARTMENTS 67
    Busted in Baraboo 69
    With Honors 70
    Twisting in the Wind 72
    White Lightning in Leland 74
    Tracers 78
    Squeeze Play 83
    A Hot Fire's Range 86
    He Can't Be! 88

COMMUNITY 95
HOW IT WORKS HERE 96
    Chief 96
WHAT EXACTLY IS SO DIFFERENT? 98
    Won't You Be My Neighbor 100

PART OF THE FABRIC **101**
    I Love A Parade 102
    What's the Buzz? 105
TRUST **107**
    No Problem! 109
THE FARM **111**
    Grounded 115
    Barred From Disclosure 117
    Healing 118

LAW ENFORCEMENT **122**
BEING A COP **123**
    A Star is Born 126
    Out of the Mouths of Babes 129
    Where's the Fire? 132
ARREST **133**
    You're Not Going Anywhere 133
    KXT996 135
    It's Always Been Ours 139
    Feigning Death 147
    Did You See the News Tonight? 148
WHO DO YOU TRUST? **151**
    Liar 152
ORIGIN AND CAUSE **154**
    Dear John 156
    Lesson Learned 157
    Right Before My Eyes 159

A-One and A-Two................................. 162

Who's At Fault................................... 165

SEARCH AND SEIZURE 176

Off Roading.................................... 177

Chasing a Bug.................................. 179

Paranoia?...................................... 186

Suspect.. 188

FORESTRY 191

REFLECTIONS IN FORESTRY 192

Branching Out.................................. 193

Poor Harold.................................... 196

Sleeping on the Job............................ 203

From a Previous Chapter........................ 205

Effective...................................... 208

Necklace....................................... 210

GITTIN' 'ER DONE 210

OPEN, SESAME! 211

As Good As It Gets............................. 212

PRESCRIBED BURNING AND OTHER FUN 214

FIRE AND RAIN 215

Firing Black Hawk.............................. 216

Let 'er Rip.................................... 217

How Could This Happen.......................... 219

What Does *Your* Daddy Do?..................... 220

When the Levee Breaks.......................... 221

EPILOGUE 229
ACKNOWLEDGMENTS 234

ILLUSTRATIONS
Photo of Author's Log House 4
Photo of Ray Larsen's Paddle. 9
Map of Fire Protection Areas 10
Framed Copy of Local Newspaper Article 17
Great Lakes Forest Fire Compact Plaque 26
Photo of Fire Control 4x4 43
Photo of Fire Control "Heavy Unit" 43
Photo of *Wisconsin State Journal* article. 51
Photo of Fire in Pine Crossing Road. 53
Photo of Tractor-Plow Furrow 54
Photo of Leland Pond. 77
Sketch of Fire Near Leland Pond. 78
Sketch of "Tracer" Fire. 80
Photo of Compromised Railroad Track at
Devil's Lake State Park 84
Photos From Don Eisberner Memorial 90-91
Author's Various Badges 124
Certified Fire Investigator Certificate 125
Sketch of Marsh Fire Scenario. 143
Retirement Plaque 229
Employee Assistance Program Retirement Letter 230
Tongue-in-Cheek Retirement Resolution 231

PREFACE

Fighting forest fires is fun. Perhaps this is not entirely true, but the alliteration proved irresistible once it occurred to me. Still, there are a few occupations that have a mystique, a hazy sort of romantic appeal. Usually, some of the appeal is genuine, but most of it is not. But my job as a wildland firefighter in Wisconsin in the eighties and nineties was well suited to me. Yes, there are the less exciting aspects of the last stage of fire suppression, the mop-up; it's tedious, it's definitively dirty, it's often lonely, and it must be done *very* well. But parts of a ranger's work are filled with anticipation, swashbuckling excitement, challenging interpersonal duels with offenders, opportunities for bestowing grace, and all in a context of considerable independence and split-second, high-stakes decision-making. There was a lot of variety in the work, from prevention through the media to pre-planning suppression in densely populated forested areas, law enforcement, and active fire suppression. It is a little boy's job, a dream of challenge, action, risk, and, most of all, the opportunity to vanquish a ferocious, heartless adversary.

As with any worthwhile activity, the fire game generates an almost endless stream of stories. The best ones include an element of learning, whether direct or in a thoughtful postmortem. Through thirty-plus years of field fire experience, I grew from the incredulity of a city boy's first exposure to the concept of a volunteer fire department to a high level of comfort with the authority associated with being a sworn law enforcement officer. It's a long path, and the most intriguing parts all have to do with people. They can be astonishingly committed, even heroic in the role they play in their community, what they'll do for their neighbors, folks with whom they share a lot line or a road, or even just a fence line. A common responsibility is a powerful unifying agent for an otherwise diverse group of people.

Still, while stories are fun, a syrupy moral attached to every one can seem trite. But sometimes there's a bona fide lesson: only a fool doesn't learn from their mistakes. So here are some exciting, insightful, and fun anecdotes. I've tried to find something to hold onto out of each account. If you don't like them, you can go start a fire and learn your own lesson.

SETTING THE STAGE

IN THE BEGINNING

The pile of firewood under the deck was solid, save for gaps to access a door and for a large window in the walk-out basement. All the corners of the pile were squared, which was not an easy task with the random shapes that branches and crotches bring to many pieces of split firewood. But I, a city kid from Chicago, had cut, split, and stacked it, and as I looked toward the house, I was quite proud of myself. My gaze drifted up. The log house confirmed the completeness of my transformation into a rural inhabitant.

A recent picture of the log house in which I lived in the early 1980s, a far cry from urban/suburban Chicago of only a few years earlier. The firewood described above was stacked around the lower door and windows.

How had I gotten here? Though I could track the specific steps sequentially, each successive step seemed to defy logic, taking me further from all that I had known in my first twenty-five years.

Four years in the corporate office of a Fortune 500 company, along with business school at night, had me primed for a productive career. Oh, but was that good enough for me? No! I decided I had to bring my business skills to bear in some area of natural resources. Why? Because I'm a child of the sixties! We not only had to do well, but we had to do good, too. So, new plan. Off I went to school at UW–Madison to study forestry and then work for a major forest products company. But in

1980, at the end of school, the private sector job market was in the tank. So much for that bright idea. I stumbled into the written test for field forester jobs with the Wisconsin Department of Natural Resources on a whim; I only heard about it the night before. I did well on the test and got an interview, and the DNR offered me a job. It paid about half of what I made in Chicago, but I wondered if it might do until a *real* job came along. Maybe I could circle back to "the plan" in a couple of years. In the meantime, there were bills to pay. I took the job!

So, there I was. The house was four rooms, five, depending on your standards, set off a country road that had almost forgotten it was asphalt. An open loft with a mattress on the floor served as our bedroom. If I slouched, I could stand in the center of that loft. But the bed took up much of the floor, so accessing the stairs that went down behind the stone chimney was just as handily accomplished on hands and knees. The setting sun shone into the living room/kitchen/dining room/family room, open to the high ceiling. Under the loft was a short hall to a bathroom and the only real room with a regular door. It served as the bedroom for our infant son, whose arrival had provided plenty of drama. More later on the birth of our son.

A three-mile drive would take us into Spring Green, population 1,265, about a quarter of the size of my high school. A village that mostly operated around agriculture, there was the usual grocery store, drug store, a couple of gas stations, and a few bars. But there was also a hardware store, a building supply store, and a farm supply store, all intersecting, to some extent, their product line to compete for the business of those in the farming trade. And there was a curmudgeonly fellow on the edge of town who could recobble almost any broken piece of farm equipment, no matter how old, with a welder. The car dealer had recently left town. Out on the highway, Doc's was the local supplier of sporting goods and liquor (funny how they go together). The gas stations still doubled as repair facilities, and there were various other fix-it shops with repair capabilities scattered through the countryside. One guy, who mostly farmed, was particularly adept at transmissions and worked on them in the lower level of his barn since he'd sold his milking herd. He did good work and for a terrific price, as long as you had three or four weeks for him to do it; repairs for a farm operation always took priority. The farther from town these places were, the more likely there was to be a price discount for cash payment.

For the most part, folks were friendly and fair and accommodating. And they usually had time to talk; those who didn't were regarded with an air of suspicion. Long lists of otherwise peculiar names in the phone book testified to the fertility of certain bloodlines. A history of German immigrants saw to the "S" section of the phone book being disproportionately large, much of it "Sch" (wipe your lower lip when you say that).

It was a time of major transition for me. I had moved from three-piece suits in the corporate world of the big city to a log house in the country with my wife and a baby and a state government job filled with trees and occasionally a little smoke.

At the Wisconsin Department of Natural Resources (DNR), perceived as a largely regulatory agency of the government, the forestry part of the work was predictable. For the most part, foresters went out, had brief discussions with property managers or landowners, and then went out in the woods and did their thing by themselves. That's why a lot of them identify this as a desirable vocation—lots of time in the woods alone. Forestry was the most respected element of the DNR, as foresters were mostly helpful and only very minimally involved in regulation.

But sea change was afoot in the forest fire side of the forestry program. Part of that change was the result of generational shifts. The older core who were in leadership in 1980 had, for the most part, a military background, either from World War II or Korea. They came from a time when choices were few, and you did what you were told, like it or not, with very little input. They brought that same attitude to the fire program either because they thought it worked or because they wanted to get their turn on the other end of the shaft. And the fire arena had a good deal in common with how military operations were run. There was a crisis, albeit of a smaller and more temporary nature, but tasks had to be accomplished, and directions had to be followed quickly and without question.

Some of the change had to do with fighting the fire itself. By its nature, there is something mysterious and other-worldly about fire. Its mesmerizing effect on the young (and less young) is acknowledged by all and appreciated by some. So the process of fighting an out-of-control wildfire was enigmatic and frightening in its own right. And those who mastered it, or at least had significant insights, took on an air of wizardry, keepers of knowledge few others possessed. That made for an exclusive brotherhood with the characteristic resistance to change such a league entails.[1]

Five years of significantly drier-than-normal weather changed everything. The years 1976–1980 saw many large and very destructive fires. The Brockway fire, in the Black River Falls area in '77, was the grandaddy, burning over 17,000 acres of central Wisconsin before it was controlled. The effects of that fire could be seen for miles on both sides of Interstate 94 for years and years, as a new crop of trees grew from the ground up. During those five years, hundreds of homes were lost, and the map of the burned area on some of the fires betrayed tactical suppression blunders.[2] The governor made appearances at the sites of some of these fires. In a program overseen by a branch of the government, such losses draw attention far up the power chain and almost always result in an urgent direction to do something—*anything*—to mollify public demand for better protection. It became clear that however the fire program had been managed in the past would no longer be good enough.

Advances in science helped enable some of that change. Improved forecasting tools and a more science-based assessment of how weather conditions and forest fuels interact brought about a better understanding of burning potential in the Wisconsin woods in spring. The days of sticking a wet finger in the air of a spring morning—"Oh, today's gonna be a barn burner, for sure!"—were over. Fire folks began to realize that commonalities existed in the management of all sorts of emergency responses: floods, civil unrest, structural catastrophes like building collapses or major explosions, wind events like tornadoes or hurricanes, as well as major forest fires. Lessons from one might thoughtfully be applied to another, and there was merit in considering organizational approaches to best bring to bear all the resources that were needed on a large fire. Beyond water and equipment, food, drinking water, fuel, emergency medical support, and a well-thought-out radio communication plan proved crucial.

---

1 In 1980, virtually all such participants were, in fact, men, though that would change in the future.

2 Sometimes the map of a fire shows it to narrow to a point, only to make another run in another direction after the passage of a front and carelessness in attending to the last hot spots from the original run. The direction from which wind blows generally shifts in a clockwise direction as a front passes, so the most careful attention is placed on the "right" side of the fire, as one stands at the origin looking down range. Failure to do so can result in a fire's "escaping" after having been contained.

Receptivity to a more thorough approach was evident in one of the exchanges at a forest fire-related training session. A bright young ranger (*not* me) complained that when he looked at the expansive burning front on a fire, he found it difficult to think clearly about how to marshal all the resources that were arriving to help him control it. The sage old ranger responded: "If looking at the fire makes you crazy, then don't look at the fire." That comment alone gave permission, even urgency, to develop a well-thought-out plan for the suppression of a fire. It takes more than just bravado.

No longer would people be hired strictly as forest rangers with no forest management responsibilities. The exigent season for forest fires was predominantly in spring after the snow melted but before everything greened up. There was important fire management work to be done during the rest of the year, but not enough to justify a full-time job devoted to nothing else. Fire management and forestry would now be integrated.

When I started in 1980, the nucleus of new DNR foresters had grown up in the decade of the sixties, or at least in its shadow. To the extent the old style of military initiation had been exercised in Wisconsin's fire program previously, it was less well received and, in fact, not essential to the operation. A forestry degree was now required, and that brought fresh ideas. There was little value to the organization in forcing an employee to be infinitely flexible for the sake of the employer just because they could. Acknowledgment of the importance of family and the needs associated with a working spouse (who often made more than a ranger) were becoming significant factors in how people made choices regarding their work.

Newly hired foresters, like me, went to a three-month-long training in Tomahawk, where much of Wisconsin's firefighting equipment was customized. Each of us figured out a place to stay during the week: a room in a local home, an apartment, a camper. The training was well done. Those teaching the fire component were a colorful group, experienced, supremely confident, occasionally profane, and generally eager to project an old-school, nothing-like-experience-and-attitude image of the fabled forest ranger.

We learned a lot there, and many relationships were developed that would often prove to be both significant and beneficial in the future. Our class, sixteen mostly younger guys, grew close. At twenty-seven,

I was fairly senior. We enjoyed a beer together at the end of the day once in a while, and Wednesday afternoons were spent in the big field to the west of the Tomahawk facility playing softball. One otherwise unlikely individual developed a well-earned reputation for being able to hit a ball over the left fielder's head, no matter how deep he played. Another, who was a skilled woodworker and the only Native American in our class, made miniature personalized canoe paddles for each of us at the end of the three months, noting our membership in that year's class of foresters. Some years later, he contracted cancer and was the first of our group that we lost. I still have that paddle.

It was 1980. The transition in the forestry program's ways and mine intersected that autumn at the end of the three-month class. Part of the onboarding process was to assign each of the new foresters to a temporary training station somewhere in the state for some practical experience in both forestry and fire control. For the most part, the stations were in the northern half of the state, where most of the fire protection areas were.[3] As mentioned, the forestry leadership took great delight in exerting their authority to assign an individual wherever they wanted to, regardless of their desires, a persistent element of "the old way." We were encouraged not to express our interest in a particular station or even in a specific area because that would almost ensure that we would be placed as far from there as possible. I don't think this was meant to be vengeful as much as it was to demonstrate the importance of doing what was asked when it was asked, without input or question.

The map shows the various DNR forest fire protection areas in Wisconsin: 1) The dark areas, toward the north and characterized as "Intensive."

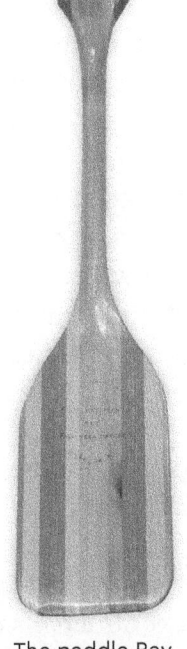

The paddle Ray Larson made for each of us in our forester/ranger training class.
It reads:
Blair Anderson
DNR Forester-Ranger
Class of '80

---

3 The fire protection program had begun in the 1910s, in the wake of the post-settlement harvest of the vast pinery in northern Wisconsin. To this day, the wildland fire problem is minimal in the flatter, mostly agricultural areas of southern, southeast, and west central Wisconsin.

In the Intensive areas, ranger stations are more frequent and generally responsible for a smaller geographic area. 2) The lighter-shaded areas are "Extensive." In Extensive areas, the geographic size each ranger station protects is larger, such as my Spring Green station. 3) The lightest areas in the west, southwest, and southeast of the state are characterized as "Cooperative." Fire departments are responsible for forest/grass fires in Cooperative areas but may request help from the DNR in the event of extreme circumstances.

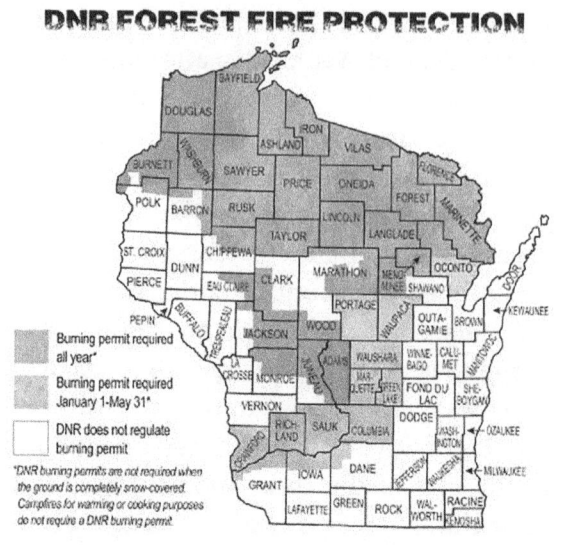

In my case, my wife and I had previously lost a child in a pregnancy that ended at the halfway point. We needed to be close to quality neonatal medical care, which, in my thinking, was Madison, where we lived when I was hired. And we were anxiously pregnant again. I didn't want to get into a power struggle, but contrary to recommended protocol, I requested a training station in the southern part of the state, adding that if I was sent to a remote part of the state, I would simply have to resign. I don't know what discussion took place behind the scenes, but my request was granted, and I was able to do both the forestry and fire control parts of my training at stations within driving distance from our home in Madison. No one had ever trained at those stations before, and I think it was with a mixture of discernment and resignation that a concession was made. Shifting priorities would soon result in additional changes.

After a few months at my training stations, a permanent opening occurred in Spring Green, a position that combined both forestry and fire control work. The supervisor for that station was the same man who supervised my training stations, and he apparently appreciated some of what I brought to the job. I was offered the Spring Green position and took it, knowing we could continue to use our medical resources in Madison, forty miles away. We moved there in the late spring of 1981, and that provided the setting for almost all the stories that follow.

It may be helpful to point out that many, even most, of these stories happened before the days of cell phones and video conferencing. I'll mention it here so I don't have to make that disclaimer every time you might otherwise think, "Well, why didn't you just whip out your phone and call?"

I changed a few names because I don't want these stories to embarrass those folks the way some of them do me.

BECOMING A FOREST RANGER

THE MYSTIQUE

Forest ranger. The mere mention of the job title transports the imagination to another world. There are romantic visions of lone advocates patrolling the mountains, communing with wildlife. It is someone supremely confident in the righteousness of their work, adamant about making the world safer for those who love the wild outdoors, no matter the risks. They live alone in a tower, always giving of themselves for the forest, going to bed each night with the conviction that in spite of the danger—no, because of the danger—there is no more sacrificial work they could be doing that bears greater benefit for the world. Forest ranger!

Now that we've got that misguided vision up on the stage, let's infuse a little reality into it. First and foremost, forest rangers always work for a government agency. All the things you've ever heard about what it's like to work for the government hold true for forest rangers, too, for the most part. The job is laden with bureaucracy, the system into which it fits tends to be guided more by rules than common sense, and there is generally little distinction in treatment between the high-performing worker and the other one.

What the job *does* have going for it, and what sets it apart from most other governmental jobs is its forest setting. Forest rangers do have offices, and there are phones in them (yes, landlines) that occasionally need answering, but the offices tend to be in remote and beautiful locations. And often, they are miles from their supervisors, which can allow a little creativity in the management of the day. When you're a forest ranger, it's easy to make a case for tending to almost any perceived need that comes up that takes you out of the office and into the field!

REALITY CHECK

As I've disclosed, I stumbled upon this fire part of the work quite by accident. The forester job, satisfying but with less of the swashbuckling allure than forest ranger, was where I settled in an interim role while I awaited the corporate mission that would meld my business and finance background with forestry. I still thought of forests and trees and logs as simply a commodity against which business principles could be applied to make commerce. I had some experience with woods forestry during various outings throughout school. But I thought of those as interesting anomalies in the pursuit of the forestry degree I would need to combine forestry and business. Still, the field forester time might look good on a resume.

But now, starting as a field forester, my perspective had to change. This new job brought about a different, more visceral connection with the woods, with the outdoors. Surely, I would be able to find ways to make some use of my accounting and finance affinities and my interest in data and statistics. But in this job, everything happens in the woods with the trees and the other plants. That new focus would require a concentration on biological specifics. As I came to grips with that, I also began to see a whole series of interconnections that I had never seriously considered before. Again, to my surprise, they fascinated me. At first, it was a series of various little factoids that stimulated me. Prickly ash, a ferocious, heavily thorned understory plant that grows in clones, is a member of the citrus family—the citrus family, in frozen Wisconsin. *Cronartium ribicola*, the fungus that causes deadly white pine blister rust, requires an alternate host, currant (genus *Ribes*). Eliminate the currant, and you will successfully eliminate the rust. As these fascinating little bits of information piled up, I came to appreciate the extent of the interconnectedness between all the different forms of life that found their home in the woods.

Trees interact with each other in various ways. When roots from one oak tree meet those of an adjacent one, they graft to one another. This can be helpful in some ways, but it also creates a pathway for the fungus that causes oak wilt, a fatal disease, to travel from one tree to the next. Black walnut secretes a substance called juglone, which is allelopathic; that is, it discourages the growth of other plants underneath its foliage, reducing competition for limited resources.

But trees mix it up with other forms of life, too, including animals, fungi, other plants, and bacteria. Abiotic factors, like weather, have an enormous effect on plant population dynamics. As I was to learn, those weather factors play heavily into the fire situation, which itself has a huge influence on plant communities, and not a uniformly negative one.

My growing awareness of the randomness of the interactions between all the creatures of the forest was liberating. It was difficult to predict the outcome of all those interactions. Still, there was a certain wild appeal in that, such a sharp contrast to the order and precision of the financial analysis with which I had grown so comfortable. I liked it.

The whole web-of-life thing suddenly became very real to me. I found it intriguing. Though something within me was still trying to angle toward the business side of things, its power was receding. Years later, a remarkable opportunity would show me how real that recession had become.

IN TRAINING

In Wisconsin, our initial instruction came through our group training in Tomahawk right after we were hired as foresters/rangers. All the newly hired foresters received the fire part of the preparation because it wouldn't be decided until later who would be going to jobs that were strictly forestry and who would have jobs that addressed both forestry and fire control responsibilities. So I was just going with the flow, surprised to find that beyond the stimulation of the field forestry aspect of the job, the whole concept of what the forest ranger role entailed was also very attractive, even though I knew it was far from my business objectives.

Some of the appeal was the tactical approach to a fire situation, especially in more dangerous weather conditions (dry, hot, and windy). But perhaps even more appealing was the sense of fraternity I witnessed among those who did most of the teaching. They were all guys. Many of them were boisterous, even borderline crude, and most of them drank, some to excess when evening rolled around. But to a person, they were profoundly enthusiastic about the fires they discussed, and it was clear that the most challenging situations they faced were also some of the most invigorating times in their lives. Adrenaline was clearly a significant part of the experience. There emerged in their stories, especially the lubricated evening versions, a thrill from a mano a mano sensation of taking on an arrogant, spirited fire. To hear the guys talk, fires seemed to have a malevolent character, even a will, that had to be faced down and defeated.

This was a whole new world for this city boy, and I experienced a bit of a rush as I listened to some of these stories. And those instructors' experiences weren't so remote; they had the same job that I might soon possibly have the opportunity to do. Contemplating the idea seemed crazy, even irresponsible. It was such a far cry from corporate work. Yet I was undeniably stirred. There is an intersection between fear and exhilaration that takes place when the chance to get involved in a romantic notion presents itself. Me? Really? Why not? Other doors were closed.

WE'RE MOVING WHERE?

A few months later, in March of 1981, a position opened up in Spring Green, as close to Madison as I could get, with its quality medical care. Our first son, Ben, had been born in December, ten weeks premature. It was six weeks before he could come home, and we knew that he might need a lot of care from people with focused specialties.

"I'm home." We were still living in Madison as all this transition worked out.

"In the back with Ben," my wife, Jan, called to me. Ben was two and a half months old, but being born ten weeks early, we were near his due date, and he'd only been home from the hospital for four weeks. His life had hung in the balance for about forty-eight hours after he was born, but he was doing well now.

"How is he?" I inquired about the new focal point of our lives.

"Another good day." Jan had taken some time off of her job after her emergency C-section in mid-December that brought Ben into the world.

"Guess what?" I offered excitedly.

We were both weary from the long adventure with Ben, and Jan had become a little leery of all the new directions my idealism was taking us. She looked me in the eye. "Don't make me guess."

Okay. She's not aware of the exciting news I have to offer, so she's not sharing my enthusiasm. I struck a pose. "I got offered the job in Spring Green."

Her serious don't-make-me-work-for-this expression persisted. "Spring Green. Where is Spring Green?"

"It's west of here, a little more than thirty miles." Actually, it was a little more than forty miles, but what I said wasn't untrue, really. I wanted this to work.

"Oh. What did you tell them?"

Good. I got to demonstrate my wise, husbandly deference to her interests a little. "Well, I told them I had to talk to you first."

"Okay. What do you think? What do you know about Spring Green?" She was in investigative mode, not puppy-dog happy like I was. Still some work to do.

"Well, it's a pretty little town along the river, small, a farm town. I've been there a few times for meetings and stuff. But here's the thing. It's the closest opportunity to Madison, so we can keep the same medical care. And I'd be working for the same guy, Marshall Ruegger." Marshall is one of the old-school fire guys, the son of a forest ranger from northern Wisconsin. And for some reason, he wanted me to fill the new vacancy in Spring Green.

After discussing it further, we jumped at the Spring Green opportunity.

THE SPRING GREEN JOB

The forestry component of the Spring Green job was to help adjacent county foresters, DNR foresters whose job was 95 percent forestry and whose area of responsibility was defined by county borders. In Sauk County, Spring Green was also within a stone's throw of Iowa, Richland, and Dane Counties. So I might be helping any of them, depending on their workload in a given year.

But the fire program took center stage for the Spring Green job. In the southernmost protection area in the state, most of the forests were

In 1997, a group of about three hundred landowners presented this to me, framed. They owned roughly four to five hundred two-acre lots in "the Pinelands" between Spring Green and Lone Rock. I had undertaken an administratively challenging effort to get a badly needed thinning accomplished without costing the landowners anything. In fact, they received nominal compensation for their limited wood. The picture is from an article in the *River Valley Home News* published in April 1981, soon after I became the Spring Green ranger. The author of the article is unknown. I was skinny, and my hair and beard were long and dark. Now ... not so much.

Blair Anderson is replacing Carolyn Agnew as Spring Green sub-district forester for the Department of Natural Resources.

hardwood, less dangerous than the more volatile pine-dominated areas in the sandy soils of central, northwest, and extreme northeast Wisconsin (more about that in the next chapter). Because Spring Green was mostly a "hardwood station," it included tremendous acreage; it was the third largest area of responsibility in the state (of fifty-eight), and the two bigger ones were on either side. It was frequently the station with the largest fire occurrence in the state, both because of its size and because it was in the more densely populated southern part of Wisconsin.

The Division of Forestry is responsible for wildland fire protection in about half of the state (see map on page 10). Elsewhere, areas of predominantly "wall-to-wall" agriculture have minimal forest fire occurrence, and what fires there are burn limited acreage and are easily handled by local resources. There are two national forests in parts of northern Wisconsin where the fire issues are "handled" by the U.S. Forest Service.

Nine fire management administrative zones with borders called "areas" of all things, covered the DNR's protection acreage. Each contained between five and eight ranger stations, generally with their own boundaries, called "sub-areas."

The fire job involved bigger program areas: fire suppression, fire department training and relations, fire prevention (Smokey!), and law enforcement. Other smaller elements varied between sub-areas. Each ranger station addressed these needs within its own boundaries, especially those that could be planned, like prevention programs and fire department training. Fire situations could necessitate assistance across boundaries and even statewide, depending on circumstances.

Many aspects of the "fire program" were more robust in the Spring Green sub-area than in any other sub-area of the state. Statewide, Spring Green involved more fire departments (twenty-two; some sub-areas had as few as two) and more schools and students than anywhere else. Students meant exciting prevention programs starring the infamous Smokey (not "the") Bear and bringing his message to the kids. My sub-area included parts of four counties, so I had to manage working relationships with four different sheriffs' departments and four different district attorneys. And fifty-five emergency fire wardens issued burning permits, more than all but one other station.

So I was in deep right from the start. And suddenly, in this work setting, from my wood-heated log house on a remote country road, the corporate world seemed far away indeed.

WHO KNEW?

I was soon to learn that there was a unique set of skills and characteristics needed on the fire side of this job. Independence, the need for initiative and creativity, high-pressure decision-making, periods of intense physical effort, and effective prioritization were crucial. For some, in strictly forestry positions, intensity and pressure were unpleasant, even intolerable. But they drew me. I discovered some strengths in myself that I hadn't before exercised. I also developed a sense of freedom and an affection for the work.

MAYBE THIS FITS OKAY

For me, the decision-making environment during fire season was very challenging but still very stimulating. Making the right tactical decision could be critical to how quickly the fire was suppressed. And that could have implications for people's safety, including firefighters, or about saving (or not saving) a barn or someone's home. In my large sub-area, it was not unusual to have two different fires occur at essentially the same time. The local fire departments responded, and based on conditions, the information provided when the fire department was paged out, and my knowledge of the fuels where the fires were reported, a fast decision was made by me, my dispatcher, and the supervisor, in some combination, as to where I would go. If a fire was in a particularly dangerous area, perhaps with lots of pine and quite a few homes, my tractor plow unit, perhaps an adjacent ranger, and maybe another fire department or two would also be dispatched to help.[4] It was always easier to turn them back after a knowledgeable fire person laid eyes on the situation than it was to try to play catch up. All fire people thought like that. No one was

---

4 A tractor plow unit is a small bulldozer (about D4 size) with a blade on the front and a DNR-designed V-plow on the back. The blade is used to move debris and clear a path for the dozer, and the plow peels away any fuels down to bare soil, roughly twelve feet wide, to help stop the spread of the fire. An associated term is "heavy unit," which includes a large water-hauling truck on a three- or five-ton frame (in the old military parlance; the frames were much heavier than three or five tons), with a pump and other equipment, capable of carrying eight hundred or one thousand gallons of water, and which pulls a large trailer carrying a tractor plow unit.

upset at being preliminarily sent and then subsequently turned back, even after they had geared up, literally and mentally, for action.

The emergency nature of fire season was exciting. The opportunity to develop leadership acumen in crucial, high-consequence situations was energizing, as was working with capable, helpful people. That included fire department folks, law enforcement personnel (sheriff's deputies, city cops, our conservation wardens, even state troopers occasionally), and the individual with the right piece of equipment, like a farmer with a disc for a fire running through a field of corn stubble. It was those folks who knew best how well I did (or didn't do) my job.

Even though wildland fires were my responsibility by law, the local fire departments were usually closer and the first to arrive at fires since my sub-area was so large. It was pretty apparent what needed to be done on most fires. Fire departments could choose to be compensated by the DNR for their wildland fire suppression efforts based on a formula (time, equipment, and personnel). But if they billed the department, the department may, in turn, bill the person who started the fire, neighbors of some sort, and that didn't always foster good relationships between tax-paying residents, on the one hand, and local government and the fire department, on the other, all of whom were neighbors of some sort. I spent time training fire departments about wildland fires and touched on what to do and the safest way to do so. In simple situations, when I would arrive, I'd seek out the chief and ask how I could help. If he was concerned or felt in over his head, he'd let me know and we'd figure it out together. Almost all my chiefs worked well that way, and they appreciated that I deferred to their situational awareness when asserting my statutory authority wouldn't be helpful.

I had to learn my way around the area I was responsible for protecting. So early on in my career, I did some self-training. I followed the advice given to the wise men after they had seen Herod: go home by a different way. State and county highways were the most efficient travel arteries, but there were many more miles of smaller town roads, most paved. That's where most people in the country lived, and that's where the fires often were. If I was going to be efficient at getting to where the need was, I had to learn those lesser roads that twisted through the hills. Neither my boss nor my dispatcher understood why it sometimes took me so long to get back to the station after a fire. Occasionally it was because I got lost. But not for long. And some of the loveliest vistas were on ridgetops on out-of-the-way town roads the efficient traveler would never see.

YOU WANT ME TO HELP WITH WHAT?

Labor-intensive DNR activities in other areas (e.g., wildlife management) drew on other disciplines, like forestry, for help. And it was the newbies who were offered up. So in my initial years, interesting opportunities arose.

I helped with the very first release of turkeys in southern Wisconsin. They were our share of a deal made with the state of Missouri, three of our grouse (which our wildlife folks trapped) for one of their turkeys. Turkeys, once native to Wisconsin, had been eliminated decades before by excessive hunting. While the prognosis was uncertain at the time, the abundance of turkeys today makes it seem unimaginable that they were ever wiped out.

Another time, the Baraboo River flooded because of a log jam of large sticks and branches, causing significant agricultural hardships. One of our wildlife guys had been an explosives expert in the Army and was tapped to make the debris go away and restore the river's natural flow. A number of us helped haul materials through flooded fields and woods to the site of the blockage, and the interstate was closed for a stretch while we did our thing. I was especially excited about photographing the explosion from a distance. I was so focused on the camera that I was unaware of the approach of a large piece of wood that landed only twenty feet from me.

Dedicated

I helped trap and tag Canada geese, retrieving them from under a rocket-launched net over a food patch and holding them by the base of the wings while a biologist sexed, vaccinated, and tagged them with a band on the leg. I lost hold of one wing on one of the birds for a few seconds. Desperate to fly away, he repeatedly flapped his wing into the side of my face, almost knocking me off my feet, before I regained my hold on him. Geese have some honorable characteristics (though a bunch of them can make a mess in a yard in short order), which I learned about in an otherwise unrelated experience.

On a return trip from a store, I came across a dead goose in the gravel next to the shoulder. Between it and a nearby pond, maybe forty feet away, was another goose, who seemed to be monitoring the situation. *Unfortunate*, I thought, but paid no further attention to it. Two days later, I happened by the same location. The dead goose was still there, but so was the live one, still close. I do not know if geese have the capacity to display forlorn expressions, but that is clearly what my eyes saw. Waiting, hoping for something positive to happen, the remaining goose had to know that its late partner was lifeless. Yet it remained. I speculated as to whether they left to eat or drink from time to time. But clearly, this faithful vigil was the central feature of its life right now. It was difficult not to reflect on how some relationships would benefit from the same sort of committed connectedness.

When I later relayed this vignette to a more knowledgeable friend, I was told that geese mate for life. That notion stirred something in me. Maybe it's the level of commitment that transcends what I think of as the "animal nature." The innate drive in a pair of animals to stay together feels as though it must come from someplace else. Such commitment surely has its advantages. But it also creates vulnerability to the pain associated with loss, as the survivor goose was experiencing. Tennyson was right.

An Injured Critter

When people learned that I worked for the DNR, I was perceived to be an authority on forestry and fire (which I became), but also on game management, water quality, hunting, prairies, all things fish, trapping, and the laws related to any of those areas. That perception was true for all DNR folks. To be fair, if someone asks a question of any of us, they should expect that contact to lead to a solution, even if it is only a quality referral. My faith in that logic got me in trouble once.

It was a Friday afternoon, late in a classic fall day, clear and cool. I was just home from work, and my family and I were about to head to Chicago to visit relatives. It was the day of landlines, and I still hadn't learned just to let the phone ring on the cusp of other plans. "Hello." There was apparently an injured animal in a field a couple of miles west of our log house. I turned to my wife, promising that a quick call would solve everything. Jan's shoulders sagged. She was wiser than I, as I learned over the years. Six unanswered calls proved the point and left me in a quandary. I might have dropped the matter, but I had told the caller, a neighbor, "I'll get it taken care of."

"This will only take a minute," I pled, but she knew better.

Off I went in our van to quickly resolve this minor issue so we could get on the road. I followed the directions I'd gotten to the field and started across the firm, close-cut alfalfa field. As I cleared a gentle ridge, I spied the quarry. An "animal," huh. Looking back at me from forty yards distance was a mature bald eagle. This wasn't quite what I had anticipated. The fierce eyes told me this was not a sick animal, and nothing looked out of place on the bird. But when I stepped out of the van, its efforts to fly were frustrated. As I slowly approached, bereft of a plan, I could see that one of its feet was preventing escape. I started talking calmly to this mighty creature, hopeful that English would convey my desire to aid. Sufficiently close, I could see that one foot, in fact, just one talon, was securely held in a spring trap. Spring traps sit rather flat on the ground, with one side of a clamp on either side of a center disc on which bait is set. Any movement on the disc releases the lightly held restraint on the clamps, and the powerful spring snaps them together. Think industrial strength mouse trap but designed to grab the leg rather than crush the neck. So, all I had to do was release the talon from the trap. Problem solved. I hadn't done any trapping, but how hard could it be?

I'd never personally considered the function of an eagle's beak. Sure, the talons were formidable. I'd seen eagles grab fish out of the water and even rabbits out of a field. But I'd never really thought about what happens next. Once it settles in a tree, or wherever, the talons simply hold the meal in place. It's the beak that literally tears it to pieces. I realized that it would take two hands, one on either side, to counter the powerful springs and open the trap. I further realized that to do that, I would have to, essentially, straddle the trap, a position presently occupied by the eagle. It was an uncomfortable thought.

I backed away a couple of steps to think. Several competing ideas ran through my mind. First, time was of the essence. As a pragmatic husband, a view of my wife tapping her foot played in the background. Second, there was a potential for strategic injury. Jan and I hoped to continue growing our family, and I didn't want this incident to negatively impact those prospects. Third, there was no plan B. If I didn't help the bird now, its luck would run out. How would I free the bird safely and quickly? A plan formed. I sometimes wore a hat of substantial leather with a wide, floppy brim. Might that be in the van? Bingo! After returning from the van with my hat, I again approached the bird, still counting on my English. "We're in this together, buddy. I want you to be free as much as you want to be free. Let's make this a win-win, huh? Whaddya say?"

When I got within a few feet, I stopped and gave him a chance to settle into my nearness. I had decided it was a male because I thought a male would better understand my potential plight in the crouched position and would, therefore, cooperate. A female, on the other hand, might have had a bad experience with a male and ... I was looking for encouragement from wherever I could find it, even such internally generated nonsense. Finally, I started to reach toward the bird with my hat, which, I hoped, once over its head, would both calm the bird and provide me with some measure of protection. Perhaps it was the pause before my final approach, but we seemed suddenly to be in sync. *He* seemed to have read my intent and stood still as I gently settled the big hat over his head. I waited a few seconds for protestations, but they didn't come. Slowly, I positioned myself over the bird when another thought hit me: *Why is there no one here to take a picture of this? No one is going to believe this.*

As I squatted to grab the opposing sides of the trap, I brushed up against the bird's body. How many people get to make physical contact with an eagle? *I can't believe I'm doing this! What a great job I have! Okay, focus, Blair. It's not over yet.* I barely began to exert pressure on the trap when the eagle pulled his talon free. He jumped up, causing me to fall backward, and started to run, the hat still over his head. A few more steps and the hat flew off, and a few more and powerful wing beats lifted him into the horizon, free again. I was aware enough to soak in the moment, taking a series of mental snapshots as the bird soared off. Everything had worked out well, and I had a sufficiently interesting story to placate the concerns my delay had caused, or at least I hoped so. Grace reigned on my return home, and the story was devoured by both little Ben and Jan, from whom the hat had been a gift.

The trouble came Monday when the warden supervisor in Dodgeville learned of my actions. From his perspective, I had blundered into a potential violation, and my actions precluded a proper investigation. Apparently, there was supposed to be an identification number on the trap, which I should have recorded, and the warden supervisor was determined to initiate disciplinary action against me. My question was, "Where were all the wardens when I tried to call them Friday night?" He was a punitive kind of guy, and I was as unconcerned with his reaction as I was baffled. His efforts withered after a few days. Nothing could convince me that I hadn't made the best decisions under the circumstances or hadn't done the right thing. I had been there with the bird, and we had a rare, shared experience and had both benefitted greatly from it. What a great job I had.

HELPING AND FREEDOM

I helped all my agency cooperators with their emergencies whenever I could. That worked out more readily with fire departments and situations closer to Spring Green. I'd respond to Spring Green ambulance calls, especially if they were south of town, where I was five minutes closer, and I knew I could have some things figured out before they arrived. A couple of times, at vehicle accidents, I was able to identify the need for a second ambulance even before the first one arrived. And I helped on structural fires when I could. Maybe that meant hauling water to a remote fire scene, as was often the need with a barn fire, or assuring embers from a structural fire didn't start a wildland fire. Sometimes I'd help direct traffic if there was a bad accident that required rerouting.

Once, I responded when the Spring Green Fire Department was paged to a fire in a historic barn at Taliesin, the Frank Lloyd Wright architecture school just south of town. Only a mile from the ranger station, I responded with my smaller truck. When I arrived, one inside wall and the inside of the roof were just getting involved, and the heat was beginning to build, a critical stage in the development of a fire. I was able to knock the fire back significantly and reduce the heat. When the fire department arrived, I got out of the way and let them finish the suppression of the fire. But the barn would have been fully involved by then had I not responded and taken action.

Being able to reciprocate with fire departments and other agencies in appreciation for all the help they provided with "my" fires was satisfying. There was some sense that all of the incidents were all of our job, and such cooperation, mutually helpful and deferential, made for good relations, even friendships. That was especially helpful in overcoming some negative general attitudes about state employees. My hope was

![Great Lakes Forest Fire Compact plaque: MANITOBA, ONTARIO, MINNESOTA, MICHIGAN, WISCONSIN — 2010 Great Lakes Forest Fire Compact Executive Award, BLAIR ANDERSON, In recognition of your dedication and long term commitment To the welfare of the compact and wildfire control]

In my later years as chief of the fire program, I had the privilege to work with an elite group, the Great Lakes Forest Fire Compact, from various states and provinces: Wisconsin, Minnesota, Manitoba, Ontario, and Michigan. We worked together on a variety of productive efforts and provided one another with mutual aid support, as spring and the fire season came earlier in the states than it did in the provinces, and our season was starting to wind down when they reached their peak activity.

to demonstrate a work ethic and willingness to cooperate that would elevate what people thought of the DNR.

Once I had eight or ten years of experience, and had demonstrated some acumen for judging fire potential, there was considerable freedom for me to make appropriate decisions on my own. In my sub-area, with my knowledge of the fuels, the location of structural developments, and the capabilities of the fire departments, I genuinely felt I knew best what needed to be done. And I was given the leeway to do things well. Rare were the days I didn't enjoy going to work.

The Dotted Line

The Spring Green wildlife manager, with whom I had become friends, had been promoted. My response was the usual mixture of joy for him because it was a good promotion and disappointment at losing a friend. Bill, the new guy, seemed like a good fellow, bright and competent, but our personalities did not draw us together.

So I was surprised when one day Bill asked me if I would have some time to go with him to get some papers signed for the renewal of an easement the state had on some private property in the Wisconsin Riverway. "Of course," I told him. But I thought, *Why do you need help getting a signature on a piece of paper?*

We made the arrangements, and when the day came, I jumped into his truck for the ten-minute ride east on Highway 14 to the home of the property owner. As we traveled, Bill began to clue me in. Apparently, this fellow had inherited the property, and he had "some challenges." He characterized the home as a little different and mentioned that there was a son in the picture whose "limitations" were far greater. Since I have a younger brother who has "challenges," that's always the direction my thinking goes when I hear about people with limitations.[5] While it wasn't yet clear to me, I understood that Bill didn't see this as a routine activity.

---

5 My brother David was born with, what they called in the fifties, severe mental retardation. When he was five years old, he was institutionalized; that was what was done then. He's remained so all his life. My only sibling, I am his guardian. He is content with his life, and he shows a sweet innocence and a pure sort of love, rare, how it is meant to be.

When we arrived at the house, I observed an older car that appeared to have suffered some in the course of its life. It struck me as a little inconsistent with what the owner of a spacious-looking two-story house and a large tract of land might drive, and my thoughts went back to the comments Bill made about the landowner. We went to the door and knocked, Bill sporting a portfolio of papers under his arm. The door opened. The man standing at the door was fiftyish, obese, in an old pair of stained khaki pants and a dirty white undershirt. Not a T-shirt, but an old-style undershirt. *The son*, I thought. I could see the limitations, and I began to understand Bill's concerns. Air wafting through the screen door had the unmistakable aroma of too many cats.

Bill opened the screen door and introduced me to the landowner. *Oh, this is the landowner*, I realized. He did not really acknowledge me but turned around and walked into the house. Bill and I followed.

This was the home of a hoarder. To say that stuff was everywhere does not begin to capture the ambiance of the place. There were piles of papers and boxes and trash, high enough that even *I* was challenged to see over some of them. A vertical pipe strapped to the wall was the only indication that a kitchen might be in one corner. We walked through narrow paths, the only accessible space in the room. Farther from the walkway, I could see where plastic bags of trash had been pitched on top of the piles above head height. Furniture occasionally peaked through the piles, having long lost their intended purpose. And all this was accompanied by the acrid smell of animal excrement.

Scattering cats signaled the approach of someone else from the next room.

"That's Stevie," Bill offered when he appeared.

"Hi, Stevie," I responded. *This* was the son. Shabbily dressed and perilously slim, he looked twenty-something, but he was hard to read. A vacant expression didn't change as he momentarily became the object of our attention. He didn't look at us, and nothing about his demeanor responded. I had an unspeakably unkind thought about the family resemblance, which I'd be embarrassed to share. What was disturbing was that he didn't interact at all; he was there but he wasn't. It was almost impossible to regard him as anything other than an object in the room. The cats were most closely tuned in to his presence. My inability to find a way to engage him felt awkward.

The landowner, in the meantime, had sat down on a chair, the only accessible one in view, at a table covered with stacks of paper and debris, except for a square foot at the edge. In that space was a bowl of what appeared to be Cheerios with milk (I wondered where that might have come from), apparently the landowner's breakfast, which he was generously sharing with a large swarm of flies. The flies calmly dispersed as he brought the spoon to the bowl, quickly returning before the spoon reached his mouth. Seeing how some people are able to live had not lost any of its ability to trouble me.

I looked up, and Stevie had quietly evaporated. I engaged the house far enough to see into the room from which he had come. At least a dozen kittens and cats scrambled in different directions. I would enter no farther.

It was clear why Bill had asked me to accompany him. Even without the most remote act of anything threatening by either of the men, I had a nonspecific but strong concern for my safety. The air was dangerously foul. However, the concern was not rooted in the men or the air but in a lack of understanding of how people could live like this and the unholy back-of-the-mind speculation of what else might be a part of such a condition. Bill obviously shared the same sense of urgency to finish our task and get out. The man signed the papers, apparently still judged competent to do so, and we left promptly.

Outside, the warm glow of the sun was restorative. Sweet, fresh air flushed the stench from my nose and lungs, and I desired to spend a few moments focused on breathing it. Adjusting back to our "normal" world, we returned to Tower Hill with the windows down, silent, except for a "Thank you" from Bill. It took much of the day to clear the cat aroma from my nasal passages. I still don't know how long the images will last.

About a year later, I learned that the landowner and his son had left the house, which had been condemned. The local fire department was asked if they wanted to burn the house as a training tool. They agreed to incinerate the house, but because of safety concerns, both structural and environmental, they declined to do any inside training. One Saturday morning, they burned the place to the ground. I don't know if any efforts were made to rescue the cats there, who likely numbered in the hundreds. I felt better knowing that dangerous living arrangement, for man and beast, was gone.

DEVIATIONS: SOMETHING'S CHANGED

"Is this the past or the future that is calling?"
—Jackson Browne, "These Days"

Despite the growing soul appeal of the ranger job, my past business aspirations continued to divert my thinking, and it took one last foray away from the woods to finally seal the deal.

Seven or eight years after I took the Spring Green position, I happened upon an ad for a job in the local weekly newspaper. Usually, such listings were for hourly labor positions at one of the local construction companies, the big glass manufacturer, one of the schools (maintenance), or one of the senior citizen facilities. The one that caught my eye was, in fact, with one of the construction companies, but this position was different, a position that would grow into the chief financial officer. Kraemer Brothers builds schools, hospitals, and so forth all over the state. At the time, they were the largest nonunion construction company in Wisconsin (and may still be).

You're kidding, I thought. A financial position with a top-notch construction company, and I'd still get to stay in rural Wisconsin? Good money, a bright future, exercising my business experience, and I could still go home to a few acres and cut wood on the weekends? This job was made for me!

So I sent them a resume, and they called. They were as stunned to find someone of my background in Spring Green as I had been to see their advertisement. "I can't believe there's someone in Spring Green, working for the government, with Fortune 500 business experience and an MBA from one of the top business schools," Kraemer's point person said on the initial call.

This is what I was up for—rural life *and* business. It all made sense. It was finally coming together. This was the job that had been tugging at my conscience since I'd walked away from the business world. I interviewed, they offered me the job, and I took it. The standard two-weeks' notice was not much for the slow-turning wheels of state, and the hastily assembled "farewell" party was more bewilderment than celebration. So I made the transition. Hadn't I?

Although the money and the prospects at Kraemer were far better than the forest ranger job, I quickly came to realize that other features weren't. There was significantly less paid time off, and because it was a more standard, year-round eight-to-five job, uniform throughout the year, there was little reason for flexibility, a feature of the DNR job I valued highly.

In the eight or so years I had worked for the DNR, I had changed more than I realized and in fundamental ways. I hadn't been at the Kraemer office for two hours on Monday before I realized that my new constitution and this new job were radically out of sync. Two weeks off a year, the same hours every day, all in the office, and inherently little variety in the nature of the work all quickly felt like a constricting box. What had I been thinking? The answer to that question bore some resemblance to information I had read in a text for a psychology class. Perhaps out of admiration for my father's business acumen and a disproportionate hankering to meet a desire I projected onto him, I had sold myself a story that I "should" be in business. That's where the money is. That's where influence grows. That's what the savvy minds pursue. I loved my dad, and I told myself that if I went in the same general business direction he did, I would be pleasing him. As is too often the case for many of us, I never really thought it through carefully, or even at all. It simply morphed into a default motivation, extraordinary in its power for having been so minimally appraised. I've since adopted the mantra many have shared with me: we shouldn't "should" on ourselves.

On Monday night, I was as anxious about my decision as I was uncertain about how to proceed, and by lunch Tuesday, I was in full panic mode, to the point that I couldn't concentrate on anything. While Jan had been happy about the change, it had mostly been because she thought it fulfilled some deep longing within me. When I came home freaked out on Tuesday, this wise woman was very calm in her assessment of things.

"If you don't want to do the job, look at your options," she counseled. That interrupted the chaotic downward spiraling of my emotions long enough to think.

I called my old boss at DNR and explained a little about what I was feeling. "We have done absolutely nothing to fill your job. In fact, the paperwork to make the position officially vacant hasn't even started through channels yet." Though I felt quite foolish, that was the out I needed. We agreed that I could come back to my job the following Monday and that the week away would go on my time sheet as vacation.

The next day, I went to my boss at Kraemer Brothers, who was also a young fellow, not much older than me, and explained things to him. He was understanding and gracious, especially considering he advocated for my hiring. "It seemed almost too good to be true that there was someone right in town to fill this job. And I wondered how you would make the transition," he smiled. I helped with some things there on Thursday and Friday, feeling both fortunate and very humble.

I returned to Tower Hill the next Monday with a dramatically greater appreciation of the job, which I never lost. I also gained discernment that helped me see people, circumstances, and situations more insightfully. I have always been thankful for that "hiccough" and the high-value, low-cost lessons I learned from it.

Even though I have been able to apply most of the tools I'd gained in the business world in some way all my life, that episode forever dispatched the burning desire I had to make a career of it. I would love to say that the deep soul satisfaction of forestry work made all the difference. That is only somewhat true. When we make choices, they change the course of our lives. Some of those changes we can foresee, but many we can't, especially as the proverbial fork in the road grows smaller in the rearview mirror. We simply can't predict the course of the path not taken, for better *or* for worse. But themes do emerge within the choices we make, and it is possible to see the beneficial path that derives from them. Is there a perfect way? Surely not. I *do* think there is value in learning to make good decisions and that we make better ones as we grow in wisdom, which doesn't come automatically. Life is hard, but it's good. I feel deeply blessed by how things worked out in my life.

A FOREST FIRE PRIMER

THE BASICS

This section is here because it's hard for me to remember that not everyone is a fire geek like I am. I have a compulsive need to explain things to the nth degree, and since it's my book, that tendency is your problem. Still, we can't take ourselves too seriously. I continue to think that understanding a few basic principles will aid in your appreciation of the stories that make up most of the rest of the book. But if it gets too tedious, be good to yourself and move on a few pages, maybe to the story about Louie. You'll like Louie. If you write to me, I'll send you a proportional refund.

All fires—wildland, structural, in your furnace, or in your car's engine, the candle on the table—need three elements to be present: fuel (grass, sticks, structural lumber, furniture, propane, gasoline), oxygen, and heat (like from a match or a spark generator). If a firefighter can devise a way to eliminate any one of those from the picture, out goes the fire.

Most rural firefighters fight either structural or wildland fires. Sometimes vehicles burn, too, but they behave more like little structural fires. Wildland and structural fires are essentially the opposite of one another in some key ways, with different fire behaviors requiring radically different approaches in tactics and equipment:

- Wildland fires have light concentrations of fuel, while structural fire fuels are generally heavily concentrated.
- Wildland fires generally generate less heat at any given location and for a relatively short time. They get hot, consume all the burnable fuel, and cool down quickly. Structural fires generate high temperatures for a sustained period in a single location.
- Wildland fires, therefore, can generally be suppressed with limited amounts of carefully applied water (or other suppressant). Structural fires need large volumes of water to reduce the high heat level.
- Burning wildland fuels are generally close to additional adjacent fuels and spread easily. Structural fuels are usually confined to isolated units, like a house.
- The key to wildland fire fighting is mobility made possible by a light water supply. Structural firefighting, on the other hand, involves setting up at a single location and maintaining an ongoing high-volume water supply.

Most fire departments tend to think about fire suppression in terms of structures. Establish a strong ongoing supply of water to pour onto the structure until it is sufficiently cooled or the fuels sufficiently depleted (burned) and fire ceases. It's like a heavyweight fight where two big bruisers stand toe-to-toe and slug it out.

Wildland firefighters (including well-trained rural fire department folks) think about fire suppression in terms of fuel patterns and opportunities to interrupt the ongoing spatial progression of the fire. If the fire is too hot to attack directly, we think roads, plowed fields, or bodies of water that will give us a chance to slip in and deliver a carefully aimed blow to just the right spot. Often, we can suppress fire as we (and the fire) are moving. Wildland firefighting can be more like guerilla warfare; look for points of weakness we can take advantage of, hit and run.

Perhaps because wildland firefighters tend to work for larger units of government (like states and the evergreen federal Forest Service) and because there have historically been so many deaths associated with wildland fires, there seems to be a tendency to make lists of rules. There are ten Standard Fire Fighting Orders. There are eighteen Watch Out Situations, which means, intuitively, that if you identify one of these situations, watch out! For those who find twenty-eight things too many to remember and still fight the fire (like me), they're all boiled down to a single comprehensive maxim: "Fight fire aggressively but provide for safety first."

Structural firefighters learn principles. But maybe because they tend to work for more local authorities, like cities and smaller municipalities, they don't seem to need such pervasive sets of rules to guide their every move.

In Wisconsin, there are significantly more fire department personnel, predominantly trained in structural firefighting, than there are wildland firefighters. So, it's usually the wildland folks who need the structural folks' help. As a result, most of the training involves wildland personnel teaching fire department personnel the different techniques associated with wildland suppression. Despite mutual teasing about which is the most dangerous, experienced structural and wildland firefighters have a strong underlying respect for one another. It's all fire.

PUTTIN' OUT THE FIRE

The actual work of putting out the wildland fire can be done either directly or indirectly. "Direct attack" involves actually putting water or dirt on the flames or physically removing the fuel immediately in front of the spreading flames. It's the swashbuckling way to think of firefighting—loud crackling flames, waves of heat in the face, the occasional singed hair.

"Indirect attack" is required when the "head fire," the running fire, is too hot, or there is a risk a gust of wind or an imminent change in fuel type could almost instantly create personal danger. An indirect attack involves removing or mitigating fuels some distance ahead of the running fire by mechanical removal of the fuel (a fire line), wetting fuels ahead of the fire, or if the conditions are just right, conducting a backfire.

A backfire is started from a firebreak, like a trail or a road, where there is a break in the fuels downwind from the fire. The fire slowly burns back into the prevailing wind to create an ever-wider area with no fuel. It is risky to fire personnel, and everyone on the fire needs to know it is happening before it starts so they don't get caught in the wrong place, like between the head fire and the backfire. A backfire requires very specific circumstances to be considered, so it is an infrequently used tactic. And timing is crucial. In thirty years (and between 1,200 and 1,500 fires), I can count on both hands the times I used backfires. When they do happen, because of the risk and desperation associated with them, the mood on the fire becomes quite intense.

WHY WAS THAT SO TOUGH TO STOP?

Many environmental factors influence how a fire burns. Most aspects of weather heavily influence changes in fuel (like fuel moisture level or temperature) and how fire can spread through them. On clear days, fuels warm in the sunshine, just as we do, better preparing them to burn. Warmer air can hold more moisture, so higher temperatures have more capacity to coax moisture out of fuels, making them more ready to ignite and burn. When that warm air is dry (low humidity), fuel moisture loss happens even more quickly. Even cool air, if it is low in moisture, can dry out fuels rapidly. And a breeze bringing fresh, dry air further enhances the tendency for fuels to liberate their moisture. Intuitively, once a fire starts, wind pushes the igniting flames and the warmer, drier air from the fire toward new fuels, further preparing them to burn.

Topography can be a factor, too. Fuels on a southerly-facing slope absorb more direct radiation from the sun, making them heat and dry out more quickly. A fire burning up a slope warms and dries the fuels ahead of it uphill, and they ignite more readily. Land contours can funnel winds through draws with steep side slopes, exacerbating the spread of the fire, and have played critical roles in most high-fatality forest fire incidents in the United States.

We have fewer large fires in Wisconsin than in the western United States because of topography. Our hills are smaller and less steep, and that means better access. In Wisconsin, there are more mapped roads, and there are many woods-access and agricultural "two-track" roads that can be utilized in the event of a fire emergency. If I can get a truck closer to the fire, my opportunity to stop or narrow the fire grows proportionally, and we are blessed with those opportunities in all but the most remote areas, like the deep forests in extreme northern Wisconsin.

Fuel characteristics, both volume and arrangement, influence a fire's spread and the difficulty in suppressing it. More fuel means more fire, and if those fuels are arrayed close together, the heat spreads efficiently, vaporizing the fuels more quickly and more completely. Think of a brush pile packed tightly so the fire spreads rapidly through the whole pile.

The Peshtigo fire in 1871 was such an inferno largely because there was so much fuel present in the forms of tops and limbs from the years-long harvest of the massive pineries in northern Wisconsin and from land clearing activities to expedite farming. And the wood in pines contains terpenes (the source of turpentine), which are highly flammable. Peshtigo was a perfect storm of conditions conducive to fire. The accumulated effect of dry long-term weather patterns, vast amounts of desiccated fuel, and a few days of weather were perfectly suited to starting and driving a fire through a perfectly set arrangement of fuels. Even the streets of Peshtigo, paved with wood chips, burned. Overshadowed by the Great Chicago Fire, which also burned on October 8, 1871, the Peshtigo fire killed as many as 2,500 people, an estimated five-fold the number lost in Chicago.

Sometimes we get ahead of ourselves, even with our best intentions. Forest fire prevention is a good example. There are few symbols better known than Smokey Bear, a young bear injured in a wildfire in New Mexico in 1950 that was rescued and rehabilitated. Smokey Bear's effectiveness has led to fewer fires. However, effective fire prevention has also caused fuel buildups that have resulted in more catastrophic fires, especially out West. Historically, naturally occurring fires, frequently initiated by dry lightning (lightning strikes not accompanied by rain, relatively rare in the Midwest), kept fuel loads down, albeit in a somewhat random fashion. Our obsession with preventing fires interrupted that natural fuel reduction process so that now when fires *do* occur, there is more fuel to burn. The fires burn hotter, with more intense and unpredictable behavior, making them exponentially more difficult and dangerous to suppress and resulting in more damage.

Our population patterns in Wisconsin also influence fire suppression activities—some good, some not so much. Wherever there's a house, there's generally a road suitable for a fire truck (or so the building codes require). That means a better opportunity to gain access to the fire. That's good. But a house in the path of a fire has to be protected, and that effort is different than suppressing a wildfire, requiring more water and, therefore, bigger and less nimble trucks, increasing risk. It also increases the complexity of the firefighting effort. One style of suppression, small and mobile, with acceptably limited water resources, is needed to fight the forest fire. Another stationary style with more significant water requirements is needed to protect the structures. Complexity is the mother of confusion.

There are also features associated with human habitation that generate risk to firefighters. Any man who's ever tried to climb *over* an electric fence can testify to its potential danger. Fuel tanks are another, potentially associated with a phenomenon called a BLEVE (boiling liquid expanding vapor explosion). Some fascinating and terrifying physics are associated with BLEVEs. Many rural homes use LP gas for heat and other gas-related needs and have large, generally five hundred-gallon tanks somewhere on their property proximal to the house. For you folks from the city, LP stands for liquefied petroleum, the country version of natural gas, and it is not a liquid at ambient temperatures. To force it into a liquid state, it is put under pressure. LP tanks are designed to handle that pressure. But if the liquid in the tank warms up, that pressure increases (tanks are always a light color so they absorb less heat from the sun). If it increases sufficiently, the tank fails. When that happens, the liquefied gas expands spectacularly, immediately growing into a vast gaseous cloud of highly flammable material. And if it was a fire that caused the heating, that cloud of gas instantly becomes a massive ball of fire with destructive power worse than your imagination. Home tanks are now fitted with pressure relief valves that may result in a flaming jet of burning fuel shooting upward from the tank during a fire. But the deadly explosions so notorious in the past are much less likely.

HOW IT WORKS IN WISCONSIN

By its nature, fighting forest fires is a stern taskmaster; circumstances impose the schedule. Fire season in Wisconsin happens every spring and is loosely recognized as starting when the snow melts, and fuels dry out, and continuing until "green-up," when the grass starts growing and the leaves come out on the trees. At that point, fuel moisture levels increase, and shade cools the woods. In southern Wisconsin, where I was, that could start as early as late January, in a dry winter, and run as late as mid-May. Those timeframes shade later as one moves north. In Spring Green, the core of fire season runs from mid-March to the end of April. Early-season fires burn in lighter fuels, which are the quickest to dry after snowmelt. That means grass fires. They start easily, burn up quickly, and can spread with incredible speed in a good wind or bad wind (depending on one's perspective), but their "flashy" fuels burn out quickly when they run out of grass. Grass fires rarely last very long.

As the season wears on and drying continues, heavier fuels begin to dry out. The occurrence of torrid short-lived grass fires transitions into less flashy but more persistent woods fires. Leaves, sticks and twigs, and eventually larger pieces of wood can ignite and burn. At that stage, a grass fire can burn to the edge of a woods and ignite those fuels. Fires in woods don't move nearly as quickly, but they burn hotter and longer because of the heavier fuels, so they are harder to put out. Even after a woods fire's progression is stopped and the advancing flames are knocked down, there is an abundance of burning material smoldering in the fire's interior. A fresh breeze can lift a hot ember and restart the fire's advance. And the longer it is dry, the larger the fuels that can become involved in the fire and the more effort it takes to put out all the burning material inside the fire line.

In a normal year, when the May rains come and things "green up," fires simply stop happening. But if the rains usually associated with May don't come, fire season can drag on into summer. Even green fuels can dry out. I recall a couple of years that were characterized as "hundred-year droughts" (I was only in my thirties?), and fires continued to occur regularly well into summer. There was one year when I worked seven-day weeks for four months with only four days off. When conditions got bad, especially when they occurred statewide, everyone was busy. There was no one to fall back on, so we just kept at it. But there was an enthusiastic fire mindset among the crazies (which included me) who enjoyed the stimulus of the ongoing work. For us, long hours and minimal days off were not as bad as they might seem, nor as we made them sound to "outsiders."

But it could cut the other way, too. Spring could be wet. As soon as things began to dry out a little, there would be another rain shower. Weather patterns would develop that would produce rain every two or three days, and they could go on for weeks, dramatically reducing fire occurrence. Now a rational person could say, "That's perfect! No fires, no being away from home. No danger. No working weekends. I'll bet those wet years are the best years, and you, as a forest ranger, are thankful, thankful, thankful when they come along." You'd think.

Our fire fraternity is an odd bunch. The fact of the matter is that wet springs induce widespread *clinical* depression.

"Ranger station," came the monotonous opening.

"Hi. This is Blair Anderson. Is Bill there?"

"Yeah, he's here. Hold on," in a flattened tone that reflected only hours left to live. The unstated message? *Of course he's here. Where else would he be? Where else would any of us be? It's spring, and it's raining. There are no fires. Today, we have nothing to do with our lives.*

So, what's the deal? Do we want there to be fires? Do we like fires? Well, no. Not really. Not a lot of fires, anyway. I guess a few are okay. After all, we need to test our equipment in real-life situations. And how do we evaluate the effectiveness of our prevention work if no one starts a fire, accidentally, anyway? And how do we assess our tactical training if we don't have a fire to test our mettle? And if there are no fires, there will be buildups of fuel, and then when it does get dry in some future year—and it will—we'll have catastrophic fuel loads and catastrophic fires with catastrophic losses! No one wants that, do they? Of course not. So, maybe a few fires would be okay, don't you think?

There's a powerful amalgamation of psychological factors that come together in a forest ranger's work. For those of us who entertain a savior mentality, the job holds a potent appeal. When there's a fire, there's an urgent need for the unique set of skills we possess. Dangers are acute. For most people, fear and confusion are pervasive. But priorities become simple. Decisions must be made quickly. Most people aren't drawn to such work, but those of us who are speak a common language and understand what seems incomprehensible to most others. It's a peculiar but tightly-knit fraternity (and it includes an increasing number of women).

Fire itself can be dangerous. However, the jeopardy can come from more overlooked sources, like smoke. There is no way to avoid taking a significant amount of smoke. That doesn't seem too scary for the fire in the grass field or the woods, "natural" fuels. But field fires that ignite brush piles at farms can be different. Often, we don't know what those piles contain, and sometimes a wind shift could put us in the smoke. Most farmers who use bagged fertilizer or pesticides burn the bags, sometimes in those brush piles. One of our younger guys contracted cancer. Though the cause couldn't be pinpointed, and he returned to work after recovering from surgery, we all thought more about the dangers, and it made us a little less cavalier in our approach to fire situations.

PREPARING FOR WINTER

There were myriad stories from the old-timers about fall fire seasons past. But as a younger ranger, I'd never experienced one. Somewhat rare, fall fire season requires a very particular sequence of events and timing. Enough sustained freezing on a fall night to kill the majority of the vegetation is the first essential. And that has to happen early enough in the fall for there to be sufficiently warm, sunny, breezy October-type days to draw most of the moisture out of the fuels. Once that drying happens, there's a good chance there'll be fires on sunny days until the snow flies.

I experienced a taste of fall fire season only once. The fire most emblematic of that fall was on a gentle west-facing, wooded hillside. There wasn't much breeze, and the fire was creeping slowly but steadily up the hill. I'd fought fire innumerable times in oak woods on slopes with heavy leaf litter. A back can or two would "normally" be sufficient.[6] But this was different. The winter's snow had never compressed these leaves. And they'd never been soaked in meltwater, under all that snow, waiting for dry spring weather to eventually desiccate them again. No, these were curly, fluffy leaves stacked deeply, compressed only by the weight of their lazy descent to the ground.

Armed only with a back can, I charged up to the most advanced point of the fire. I'd knock out one side with this can, refill, and get the other half with a second can full. I began my attack, accustomed to a squirt or two being sufficient to subdue a wide swath of such a fire. Not today. The ineffectiveness of my work surprised me. Two more sprays were helpful but not conclusive. The leaves were so fluffy that the fire underneath was effectively shielded from my extinguishment efforts. This was going to take more than I had expected—more water, more time, more effort. With only twenty feet or so of fire line suppressed, I realized I had already used a significant part of my water. And I was discouraged to look back and see that two spots in my short fire line had flared back. Suddenly, I had a sense of hopelessness, and realistically so. Fortunately, the fire department had found a path to the nearby top of the hill, where the woods morphed into a cut hayfield. With pumps, hoses, and trucks full of water, it was an easy place to stop it.

I had a new appreciation for some of the old stories I'd heard. And now I had one of my own.

TIME TO GO TO WORK

Fighting a fire is, in many ways, the ultimate challenge. Beyond the broad caveat of safety, there really are no always-do-it-this-way rules. An individual is usually responsible for starting the fire, but that is a matter to be determined through the investigation *after* the fire is arrested. There are principles about fighting fire, but all fires have different combinations of characteristics: different fuels, topography, current

---

[6] A back can is a container of water, carried on the back via straps, with a three-foot hose, leading from the base of the container to a manually (two hands) operated brass trombone pump with a nozzle. Utilizing a pair of one-way valves, the pump is extended, drawing water from the container, then retracted, forcing the water out of the nozzle, like a heavy-duty super soaker. They weigh about fifty pounds full. The newer cans were galvanized, and some of the older ones were brass. Later, canvas bladder bags became the norm. They are more comfortable to wear but have a slightly smaller capacity than the old five-gallon cans. Sometimes, that little extra water makes a big difference.

weather conditions, approaching weather patterns, opportunities to attack the fire, obstacles (a road, a river), exposures (houses, barns), and resource availability. So the decision-making that takes place on a fire is never the same, and one can only be reasonably called out for bad

A photo of a modern Ranger 4 × 4 fire vehicle. The units I used were similar in design, though early on, the pump was under the hood and operated off of the engine rather than a separate gas-powered pump, as is the current design. Water capacity is roughly 150 gallons.

A modern heavy unit, similar to those in the past, consisting of three components: a three-ton or five-ton truck (using old military designation; the frames were much heavier) carrying between eight hundred and one thousand gallons of water, a pump, and enough tools for a six-person crew; a trailer; and a crawler tractor (or "tractor plow"), equipped with a blade on the front and a custom-designed plow on the back. In my day, these were usually John Deere 450s. After 1982, the hydraulic controls for the blade and plow were configured identically in all such units (about eighty around the state), as discussed in the story "He Can't Be!"

decision-making when it is blatantly apparent. Any tactical decision that stops the fire, as long as it doesn't push *too* hard against any of the principles, is borderline heroic.

If it was dry, you worked, no matter what day it was; I worked far more Easters than I didn't. And depending on the weather, you worked late, sometimes even if there weren't fires, because conditions were right for an ignition, or your neighboring ranger was tied up with a fire and unable to respond to an additional fire that might pop up elsewhere in his area. Normally, work ended at four thirty, but we'd stay at the stations as late as eight o'clock at night in fire-favorable weather in case something should start.

Sometimes we would "drift" or stage to a different location. That might be because a neighboring ranger was going to a fire near the far edge of his protection area and might need help, either with that fire or if a second fire occurred elsewhere. Drifting could generate better spacing for fire units not already committed to a fire. The idea behind "drift" was just like it sounds: cruise in a general direction, a nonspecific destination. We'd settle into a nearby town, at the fire station, or at a wayside along the highway, and wait to see what developed.

When fire season was busy, we would sometimes go home on "standby," which meant we had to be ready to respond to a fire at a moment's notice.[7] That imposed certain restrictions, the most important of which, to some, was no alcohol. Occasionally, when we closed stations, there was an unspoken conjecture that the boss didn't put us on standby when he should have. That could generate some inter-ranger discussion at the end of the day.

"Hello."

"Hey, Blair. It's Ralph. Sounds like you guys were fairly quiet today."

"Yeah, just a couple. And they were easy. But the one Sauk ran on could have turned into something interesting if they'd gotten there two minutes later. I thought today would be busier."

"Yeah. We were quieter than I expected. But it's dry out there. I was a little surprised we aren't on standby."

---

7 Standby was a work status designated for an evening or, occasionally, a weekend day when the weather was not deemed sufficiently dangerous to work, but too much so to be fully liberated. As I recall, we received an hour's pay for a standby shift, which was eight hours. During fire season, we'd typically go home at the end of the day, whenever that was, on standby. That was okay. But weekend standby kept folks from going anywhere with their families, so it wasn't used often; either we worked or we were completely off.

"Me, too. The humidities and winds this evening seem right for fires when the six o'clock burning permits kick in. I'm gonna stay ready."

"Same here. I'll bet you a quarter one of us has something tonight."

"I'm not taking that bet. I think you're right."

And so we, in a sense, put ourselves on standby, but without the pay. But those evenings were rare. And the disenchantment usually filtered back to the boss, either directly or when we had a fire or two or three on such an evening. And our supervisor was a seasoned ranger, so it was never an us-versus-them thing. Relationships were good, and that mattered.

In fire season, I always had a pager or a radio with me, and if one of "my" fire departments was dispatched to a wildland fire that sounded (to me) like it had dangerous potential, I asked Jan to call Louie, my dispatcher, at home, and I immediately went to the station and responded with my truck; no one had to "send" me. That was one of the benefits I gained with experience. If the fire weather was especially bad, or if I didn't get done until late, or if I was going to have to check a fire at first light the next morning, I would take the truck home with me. On short nights, the twenty minutes it would take to drive the truck to the station and then my car back home made a difference.

Louie

Louie, our dispatcher, was a key person in our fire system who plays a role in many of my stories. A dispatcher gathers information and intelligence and conveys it to the agency's resources to address their responsibilities. They also track and assist those "field" resources as they do their work. When you call 911, you speak with a dispatcher, who sends whatever you need to wherever you need it. That's what Louie did for us. But he was more than just an information processor. He was our link to reality on some days, a source of reminders when the day got too long for us to remember an appointment or a meeting, and a general mother hen who always made sure that we were accounted for. If we were at the scene of an incident and we didn't check in for too long of a time, Louie's resonant, silky-smooth voice would calmly come across the radio: "Spring Green ranger. Dodgeville. Status check," just to make sure we were safe. His cool temperament was a balm. It was the same at eleven in the morning as it was at eleven at night. Sometimes in the spring, if the days (and nights) were busy, he seemed to speak with my wife on the phone more than I did in person. He answered logistical questions, looking up information on people with whom we were in contact, sometimes because we asked and sometimes on his own initiative. Louie was a reassuring and encouraging voice in the night. His meticulously accurate dispatch records were crucial to the reliability of our investigative activities and reports.

Louie was, is, a friend, and we were blessed to have him watching out for us.

Up In Smoke

Things had greened up, and there hadn't been a fire in a week or so. We all thought the season was done and had shifted our mental focus to other duties. So, I was surprised to hear the pager go off for a grass fire northeast of Spring Green. I responded, falling in right behind the Spring Green Fire Department on Highway 23 to reach the scene of the fire.

It was up in a valley, betrayed by a low, lazy haze. The vegetation in the woods was green and already a foot or so tall. Flame advancement could only be detected by the generation of smoke hidden low in the grass. Green fuels burn coolly, slowly, combusting incompletely, resulting in a lot of smoke, as most of the heat is consumed, evaporating the moisture in the green plants. Fire had crept away from a small brush pile, the product of post-winter cleanup along the hayfield's edge. It was creeping leisurely up the gently sloping hillside through the woods adjacent to the field. It was an easy and fast suppression, wading through ground-hugging smoke, and within a few minutes we were loading stuff back on the vehicles, ready to head back. That was the end of that.

But a couple of days later, I found myself scratching my right leg repeatedly, just below the knee. I didn't think anything of it until I realized later that I was doing the same thing with my left leg. Upon inspection, I found the beginnings of a rash. By the next day, the rash was fully developed, but it had some unusual characteristics. It proceeded only a couple of inches below my knees, at which point it abruptly stopped, almost as if someone had drawn a line around both of my calves. It continued up my thighs, as well, but faded out quickly. The itching and subsequent development of watery blisters made poison ivy an easy diagnosis. It was not a stretch to believe that there was poison ivy among the vegetation that burned in the recent fire.

The agent that causes poison ivy is an oil in the plant called urushiol. I knew the oil was present in leaves and stems, even the roots, and could be problematic even in winter if one encountered it clearing brush. What I learned from this fire is that urushiol continues to impose its effect even when the plant is burned; it becomes part of the smoke, too.

As I thought back to the day of the fire, I remembered that the smoke hung very close to the ground, such that its impact didn't go very far up my legs. I also recalled that I had been wearing some long socks that came most of the way up my calves, resulting in the peculiar pattern of the rash. As I thought further, I realized how fortunate we had all been that the smoke didn't rise such that we were breathing it. I don't know what the impact of it inhaled would have been, but I didn't want to think too much about that, and I had a particular appreciation for how the rash faded up my legs.

IT'S AS SIMPLE AS THAT

Spring always broke down work planning to the simplest of terms. If the snow had melted and the sun was shining, the activity of the day was fire vigilance. Period. Everything at work and everything at home, barring very urgent matters, took a backseat. It wonderfully uncluttered the day's to-do list. Whatever the request was, the answer was no if it wasn't fire. For that day, only one thing mattered, and to be done properly, it required one's undivided focus. Other lesser activities were performed, to be sure, but the premise had to be that it could be dropped at a moment's notice. All eyes were on fire. If your response to a fire was delayed by something else, there was no excuse.

In Sauk County, where most of my area of responsibility lay, I had worked out a deal with the sheriff, who did all the paging in the county. I had my own paging frequency, and one of their dispatchers would page me whenever they sent any fire department in the county on a wildfire. I could hear the pages for most of the fire departments directly, but not all of them. I got to the point that I recognized the first tone of our page sequence so well that on bad fire days, I would sometimes be out to the truck and moving before the tone finished and the dispatcher began their spoken message. On those days, everyone was in the starting blocks for the next fire.

There were benefits to the need to be unshackled from other responsibilities.

"Why, yes, I'd love to be part of that meeting (or work activity or other gathering). But it's fire season, so I have to stay close to the ranger station and be ready to respond to a fire. Sorry." If another agency was conducting a controlled burn in mid to late morning, I could be justifiably on the scene since it was me who issued the burning permit. I could be around the flame and in the smoke and part of the action, but without any responsibility for the fire since I had to be free to leave in case a wildfire broke out. All the fun and none of the accountability. It was a wonderful feature of such a spring day.

The extreme demands of the ranger's job in spring were great. But the comprehensive freedom from almost any other responsibility (like meetings, ugh!) was like a kid's life. And we got paid for it!

SOOOO BIG!

When I spoke with school kids about forest fires, they were invariably fascinated. Aside from my large physical presence, the badge and the brightly-colored fire attire (a bright yellow shirt on me and bright orange coveralls on Larry, my equipment operator who managed the heavy unit) were different and dramatic, and the kids were drawn to it. Young kids have the ability and a willingness to treat people like heroes. It's too bad that that fades with age.

The lavish discussion of heat and flames and saving trees and animals and people's homes always had the kids wide eyed and focused. But invariably, we reached a place of reflection when we talked about big fires. That time came when we got to the issue of how it started and, more to the point, how big it was when it started.

"It's really scary to think about such a big fire, isn't it?" The language was scaled for the age of the particular class.

"Oh, yes!" the young ones responded. By sixth grade, there were many to whom it was important not to be impressed by anything.

"But how big do you think that fire was when it started?" That was the time when the looks of boundless enthusiasm morphed into the furrowed brow and the clear message that they had reached a disconnect. You see, the idea that such a raging inferno began with a flame that any of the kids could have easily blown out didn't seem possible.

At that point, I would produce a book of matches, making appropriate disclaimers as to their use, and light one. "Most fires are only this big when they begin."

As I moved forward in time, from the little flame on the match to the piece of newspaper, to the small material in the brush pile to the two-inch sticks to the larger pieces, and then to the grass a few feet away, and across the field and into the woods and up the hill and into the pines and cross country, some kids started to make the connection.

"I'll bet all of you could blow out this match, couldn't you?"

"Oh, yes," came the unified response.

My message was prevention, and it was a sound one: fires are easiest to control when they are small. And some of the kids got it. And for some of them, my message was part of their what-did-you-learn-at-school-today response when they got home. Maybe that did some good, too?

I lived in Spring Green long enough to know young adults who had been part of my presentations when they were little. More than one said they still remembered the message.

Squeaky

One spring morning, I got a phone call at the office.

"Blair Anderson."

"Hi, this is Steve Hopkins." Steve Hopkins. I knew the name. He was the outdoor editor at the *Wisconsin State Journal*, a Madison-based newspaper. He was older, or it seemed so when I was younger, well into his fifties, almost my father's age. I had seen him interviewed once; there was a disarming easiness about him. "I'm with the *State Journal*, and I'd like to do a piece about your work."

Me? My work? Really? "Oh," I came back smartly. Pause. I think he was busy being impressed with me.

"I thought maybe I could come out there and spend some time with you during fire season."

"Uh, okay. What specifically did you have in mind?"

"Well, I thought I might come out and ride with you for part of the day. Would that work?" he asked brightly.

"Yes, that should work. There are some things in the field I can do, and we can talk as we ride. But I must warn you, sometimes I can get committed on a fire or go from one fire to another and not get back to the station for a very long time. So, you could be stuck with me for a while. Is that okay?"

Like a good reporter, he quickly came back, "Oh, yes. I understand. That's how fire season is sometimes, right? In fact, that would be interesting." These days, there would likely be some red tape to get approval for having someone who's not a certified and trained firefighter riding in the truck. But back then, we were a little closer to the Wild West, and the lawyers' concerns with liability weren't so intrusive.

So, we set a date, and he came out to Spring Green, and we spent several hours together. As it worked out, there weren't any fires for me to respond to during our time together. But I did investigate one from the previous day. I was able to explain some of the indicators we look for on a fire to try to piece together what happened, and he took a few pictures.

We also stopped to visit one of my fire wardens, who happened to operate a small family-owned cheese factory. There used to be a lot of those, and they made excellent cheese. But over the years, most of those little family operations closed in favor of bigger, larger production facilities. After a discussion with my fire warden about a prior situation, he asked if we wanted some fresh cheese curds. That is an offer to which I would never say no, so we bought a bag and went on our way. Now, you may or may not know much about cheese or cheese curds, so I'll fill you in. Cheese curds are freshly produced after the initial curdling of fresh milk, the first step in making cheese. If you're lucky, and your timing is right, you can get them when they are still warm from the initial process. That day, good fortune was ours.

WHEN THE SMOKE CLEARS

Forester-ranger sees light at end of firefighting tunnel

By Steve Hopkins
Wisconsin State Journal

It is nearing midmorning when Blair Anderson leaves the ranger station at Tower Hill State Park, crosses the Highway 23 bridge over the Wisconsin River at Spring Green and aims his truck for the hills of northwestern Sauk County.

Anderson is a Department of Natural Resources forester-ranger. He and Larry Severtson, a forest fire control assistant, are a two-member firefighting team working out of what is known as the Spring Green sub-area. The sub-area includes almost all of Sauk County, the northern half of Iowa County and pieces of Dane and Richland counties. It is part of the Dodgeville area that includes a big chunk of southwestern Wisconsin.

Anderson will spend most of the day visiting old burns, checking records in local fire stations, making out reports. It is a luxury he has not had until now, a luxury provided by Monday night's thunderstorm. He will check out two fires, small ones started by lightning during the storm. Both were handled by local volunteer fire departments.

Anderson's truck is a bright yellow, 1-ton pickup rigged for firefighting. It carries 150 gallons of water, portable water cans for backpacking into burning areas, axes, shovels and chain saws.

Anderson stays in radio contact with Louie Weston, the fire dispatcher out of Dodgeville. Weston reports nine lightning fires burning in Marquette County, near Westfield. All appear to be small. All appear to be under control. Marquette County had Monday night's lightning without much of the accompanying rain.

"It's good to have Louie there when we're on a big fire," Anderson comments. "He's sort of like Radar used to be on the old M*A*S*H television show — he knows what we need even before we ask for it. He's comforting to have on the job. He's our calm voice in the night."

Anderson scans the countryside from the cab and sees green where the day before there was no green. "The bad part," he says, "should be over. Every time it rains, every time things get a little greener, the danger of fire is lessened."

Between Jan. 1 and midmorning Monday there were 753 fires in the state, burning a total of 2,161 acres. About 40 of those fires were in the Spring Green sub-area.

Most in Anderson's area were small brush fires. One day he had four fires going at the same time. The largest was two weeks ago near Bear Valley. It was a 100-acre, wind-driven blaze that swept out of control up a valley. It climbed a hillside, burned across the top and was headed down the other side before it was contained by men carrying back cannisters who had climbed up the back of the hill to meet it.

There were about 80 volunteers fighting that one. Firefighters from Plain, Loganville and Lone Rock were on it. Volunteers from the area showed up to help. It was started by a backfire from an all-terrain vehicle.

The scars there still are plainly visible — the burned-over hillside, the deep furrow dug by Severtson's tractor-plow, a piece of equipment armed with a blade and a cutting disc that will tear up roots, move rocks and dig a deep furrow in just about any piece of ground you put it up against.

The land will recover. There are green shoots poking up even now through the burned and blackened earth. Life goes on.

Anderson, 36, is a transplanted city boy. He grew up in Chicago, earned a master's degree in business administration from Northwestern University and spent four years working for a Fortune 500 company in the Chicago area.

He felt confined. He longed for the wide-open spaces. He wanted to apply his business experience to the natural resources area. He went for another master's degree, this time in forestry, from the UW-Madison. He has been the Spring Green area forester-ranger for eight years.

Anderson's territory falls in the category of an extensive protection area, which means that local volunteer fire departments make the first response to fire calls. In an intensive area, in the often tinder-dry piney woods to the north, the DNR, with larger firefighting units, makes the initial response.

There are 20 volunteer fire departments in Anderson's area. He keeps in touch with all of them.

On this day he stops at the Kraemer and Sons Construction Co. Plain, to confer with Plain Fire Chief Bob Neubeisel. He checks in at fire stations in Loganville, Reedsburg and LaValle. He stops in Ironton to talk to Village President Pat Mortimer about a recent brush fire there.

He stops at the Schmitz brothers' cheese factory in Bear Valley to check in with Erwin Schmitz, a regional fire warden. He buys a pound of fresh curds from Joe Schmitz and

Ranger Blair Anderson examines lightning-struck tree near Loganville.
State Journal photo/STEVE HOPKINS

munches on them happily as he drives away. "I love them when they squeak," he says.

Near Loganville, he leaves the road and follows a set of tracks along a fence line to a lone ancient oak that was struck by lightning the night before. There is a rip in the tree. A long, narrow burn runs along the fence row. He picks up a post, inspects it. He pokes in the duff at the edge of the burn, the dried leaves and dead grass.

He uses the generic term — "fuel" — for all these things. "Sometimes," he says, they will smolder and break out again later."

His job is part fire prevention. Burning permits still are required and will be until the end of May. He doesn't like to, he says, but he will issue citations for illegal burning. "If I don't, it will get out of hand," he says.

He stops at radio station WRDB on the edge of Reedsburg where he sometimes does radio spots and news interviews relating to fire safety. The station manager is out. He scribbles a note and leaves it on her desk.

It is not quite 5 p.m. when he pulls back into the Tower Hill ranger station. He is in a good mood. More rain is predicted. He sees the time nearing when he will become more forester than fire ranger again.

But it is not over yet. There will be two more fires that night — near Loganville and near Arena, in opposite ends of the county — both started from lightning-struck trees that had been smoldering all day. He will be out again until near midnight.

He still is optimistic. The bad season is nearing an end. More rain is predicted. The signs are good.

Steve Hopkins article about the author from the *Wisconsin State Journal*, April 28, 1989.

"I love them when they're still warm," I reflected out loud to Steve. "They squeak when you bite into them." Of all the things we talked about that day, that remark is one that made his article.

It was well written, accurate, and featured a photo, all very folksy. It was fun to be in the paper, but I didn't really appreciate the reach or the impact of the newspaper until a week or so later when I received in the mail two clippings of the article from two residents. They were acquaintances, but I didn't know them well enough then to call them friends. Still, they had taken the time and stamp money to send me the article with a brief note of thanks. Even more, I received a 9 x 12 envelope in the mail a couple of days later. In it was a laminated copy of the same article with a more thorough expression of appreciation for the work I do in the community. This came from a similarly distant connection. I was touched by these expressions and had a new appreciation of the extent of my profile in the community. People pay attention.

Take Cover!

It was April, and we were in a prolonged dry stretch. Even the larger fuels on the forest floor were getting dry enough to burn, making mop-up on any fire in the woods very time-consuming. Anything in the Wisconsin River valley, in the pines, would require the efforts of all our local rangers, including the new guy in Richland Center. So, on a day with a healthy west wind, when Grant County paged out Muscoda Fire Department to a fire in the river bottom a couple of miles west of Muscoda, pine country, everybody bolted for their trucks.

The village of Muscoda had enough yards and fuel-free open space that the village itself was probably not in danger. But the pine stands had been planted right to the edge of the village, and there were a lot of houses that had been constructed in there. They were susceptible. Sometimes, tree branches grow, so they reach the house or even lie on the roof, causing a dangerous fire situation. We had done exercises in the area and knew the potential for destruction. The fire was in the Boscobel ranger's fire area, but the Richland Center ranger, my heavy unit, and I were dispatched immediately, along with a call for four additional fire

WHEN THE SMOKE CLEARS

departments, should significant structural protection be needed. This was a fire we had planned for.

What I didn't know was that the Boscobel ranger, who had more experience than I did, was unable to work that day because of a family illness. That left me as the most seasoned ranger on the fire by a long shot. As I drove to the fire, monitoring radio traffic, I realized that I would probably be the one making, or at least endorsing, the tactical decisions we made on the fire, a daunting task, especially with so much at stake. An additional ranger was dispatched to the fire.

When we arrived, I saw that the fire was largely in an area where there had been recent timber harvesting. So there were tree tops and branches in a twisted mess—lots of fuel and slow going for our equipment. I had Larry begin putting in a fire break close to the right edge of the fire, starting at the road from which

Another picture from the Cottonville fire shows the advancing flames crossing a road. Fires burning in pine fuels in such dry and windy conditions display extreme and erratic fire behavior. They generally cannot be stopped until there is a break in conditions—a lull in the wind, a break in the fuels, or the cooler temperatures and higher humidities that begin to appear late in the day. We had fifteen miles of control line all the way around this 3,400-acre fire by two o'clock in the morning. It held during the next day's similar weather. Picture courtesy of Chris Klahn/Mike Lehman.

the fire began. He would need to use the blade on the front of the tractor plow to clear a path and put in a furrow, so it would take him a long time to make progress. The chief of one of the departments responding to the fire had experience with this sort of fire, and we were comfortable putting him in charge of coordinating the fire department's efforts to protect structures. We had developed structural maps of the area, showing where structures were, their fire numbers, and roughly how far off of the road they were. This information might seem like it would be apparent, but for someone not from the area, and with smoke, fire, and panic in the air, such maps were crucial for effective response.

The other two rangers and I got together to consider where we might make a stand. We were fortunate that there was a quarter-mile stretch ahead of the fire without any houses. Since most of the fuels were on the ground, the fire was very hot but advancing slowly. That gave us some time to plan. We identified an area where an underground gas line had been installed and which the gas company kept clear of vegetation so they could monitor their line. That right-of-way was twenty to thirty feet wide and

The furrow that the plow on a tractor creates is nearly ten feet wide, a good break of mineral soil that won't burn.

was clear of anything except mowed grass. It ran perpendicular to the direction of the fire, the perfect place to try to make a stop.

The left side of the fire bumped up against a powerline right-of-way, also with minimal vegetation, and some fire department units were able to contain its lateral spread there. Beyond the gas line right-of-way, to the east, is where the structures began. If we could not hold the gas line, three structures would be threatened immediately, and four or five more would be threatened within the next ten minutes, a large assignment, even for four or five fire departments. We only had ten or fifteen minutes before the fire would reach our position on the gas line.

The burning materials were putting up ten- to twelve-foot flames. The standing trees were sufficiently scattered that they could not sustain a running crown fire, but they torched dramatically when the fire reached them.[8] With the heat from the fire, there was no way we would be able to stop the fire directly from our position. The idea of backfiring from the right-of-way quickly occurred to me, and I suggested it. How good an idea it may have been was questionable, but there was nothing to lose. We hastily communicated our plan to the other units on the fire, positioned our vehicles on the right-of-way, and began lighting the backfire.

Aaron, the new ranger from Richland Center, lit the edge moving toward the north, toward the powerline right-of-way. The other ranger lit toward the road to the south, starting a break that we hoped our tractor plow would eventually meet so we could pinch off the fire. As the backfire burned, it made what seemed like painstakingly slow progress. Flames from the main fire advancing in our direction were coming into view. Time was tight. Aaron continued lighting the edge to the north, stopping regularly to ensure that the fire ignited sufficiently. As he went, he got farther away from the nearest truck, and I began to grow

---

[8] A running crown fire is a fire that is driven by the wind through the tops of trees, independent of the fire on the ground. It almost exclusively occurs in conifers, usually pines; in Wisconsin, that's red pine or jack pine. It is considered extreme fire behavior and most often occurs under prolonged windy and very dry conditions. A running crown fire is impossible to stop. The suppression tactic is to stay close to the fire as it progresses, waiting for a break in fuels or a break in the wind, when it drops out of the crowns of the trees and can be suppressed.

concerned as the head fire advanced toward us. He was getting close to his objective, the powerline right-of-way, but I didn't think he sensed the urgency; he was doing a careful job, but this situation called for a fast one. I felt responsible for him, and I was increasingly anxious for him to finish and get back to the safety of the trucks. I hollered to him, but between the noise of our engines, the whining of the fire department truck sirens, and the hissing and crackling of the fire, there was no way he could hear me. I glanced toward the fire again; we had a minute. I ran toward his position. He never heard me until I reached him, and he startled when I yelled at him to get back to the truck.

As we raced back to the trucks, the main fire was beginning to have a drawing effect on our backfire, just as we hoped, pulling it against the wind, drawing it into additional fuels, and widening the break. Aaron and I weren't back to the trucks for thirty seconds when the center of our backfire, which we ignited first, got sucked back up into the head fire's embrace. There was an explosion of fire, and it continued up and down the fire line like a breaking wave. An incongruous silence quickly followed as the flash fuels were all consumed. A couple of embers dropped behind the gas line and ignited fires. Our focus was there, and we pounced on them quickly while they were still small and were able to stop them all. As dramatic as the two fires colliding had been, the sudden end was just as stunning. We had stopped the fire and saved the houses.

There was still much more work to be done with mop-up, and we had to stay vigilant, but we had no further problems. I apologized to Aaron for startling him, but he appreciated my concern. He admitted that he was surprised at how quickly the fire intensified. Cheap lessons are always the best.

Night Fires in a Marsh

"Sometimes the road leads to dark places.
Sometimes the darkness is your friend."
—Bruce Cockburn, "Pacing the Cage"

Some ways of putting out a slow creeping fire in a marsh at night work better than others.

It was almost nine o'clock at night when the pager went off. On arrival, a line of fire, maybe three hundred yards long, was burning at one end of a long strip of marshland covering five or six hundred acres. The fire line was in two stretches, separated, I surmised, by a patch of open water, essentially two separate fires. I had sent a small contingency of the responding fire department to the far side. That side looked longer than the other and was accessible from the highway. I told them I'd take the shorter side, which I thought could be reached by a two-track road, essentially a glorified trail hunters used to get to the back of the marsh, closer to the river. I told the fire department that as soon as they finished their side of the fire, they could head back to their trucks, back to the station, back home to bed. That's always a good motivator for tired folks in the night.

There is a lovely luxury associated with a fire like this in a marsh on a cool night. Unlike almost any other fire setting, there is essentially no urgency. The fire is a long way from any structures, and with dropping temperatures, the flames are spending most of their energy driving off the moisture in the air, with little left to burn the grass. On this fire, I could have stamped much of it out with my boots. It may well have burned itself out in the cool, humid night ahead. And the marsh grass itself was of no value where it was. In fact, the marsh would grow greener that spring after being liberated from all the matted dead growth from the previous years.

So, when I arrived at the leading edge of the fire, I decided to take a minute to enjoy it. On a clear, moonless night, several miles from town, this silent ribbon of fire was the only source of light, save a few emergency flashers and an auxiliary flood light from the distant fire trucks on the highway. I added to the peace of the scenario by shutting off my truck and flashers. In such an isolated and quiet setting, fire takes on more the role of comforting friend than ravaging foe.

After enjoying this long enough to see the opposite fire line shorten from my fire department friends' efforts, I got a call from the chief inquiring about my status.

"You okay back there? I'm not seein' your lights."

"Yes, all good, thanks. Just starting in on the fire," which was true, sort of. "This won't take long. Looks like you're making good progress. Take off when you're done. I'll be good here."

"Ten-four."

The rest of this would be my own thing. The fire department was situationally released. My dispatch was down for the night, though I'd need to "check out in the blind" when I was done so Louie would hear and be assured of my safety and could move from the fitful sort of sleep he slept when any of us were out at night, to REM. The sheriff's dispatch was up, and I could call them if I needed help. I casually strapped on a back can, ready to start putting out the fire from the road's edge.

But one thing first. My dad didn't completely understand all that I did. His wife, my mother, had died unexpectedly several years before. He was alone now, and the rules of communication between us had changed. So, on that beautiful starry night, in the middle of a pitch-black marsh, I called him from the bag phone in the truck. He had been asleep, just barely, but he was comforted to hear from me. He enjoyed hearing about where I was and what I was doing, and perhaps most of all, that I thought of him at that moment. We spoke only briefly, but it was the perfect entree to my suppression work.

Without the urgency, without the adrenaline, there was something quite pleasant about this stroll, alone, under the night sky, with a softly crackling set of night lights to guide me on my way. *What a great job I have*, I thought. Under these conditions, the fire was easy to suppress, and I maneuvered around the wetter spots with the aid of the fire's soft glow. (You're probably already anticipating what hadn't yet occurred to me.) I was able to conserve my water, sure that one can would be more than sufficient to finish the job.

And it was. Partway through the task, the chief called again to say they were leaving, and I was left alone in the marsh. I stopped occasionally to stretch my shoulders and admire my work. Near the end, I could see some faint reflection of open water, which marked the end of the fire. I "launched" a few last squirts in that direction, not wanting to go ankle-deep in muck, and it was over. Satisfied with what I had accomplished, I looked up at the night stars, pleased with my lot in life. Time to head back.

You know what's next, don't you? The glow of the call to my dad and of the fire were both gone. The first few steps back toward the truck were obvious. But as I looked up toward the truck, I couldn't see it. Maybe it was more to the right of where I was looking. Or to the left. Okay, I didn't know for sure where the truck was. But I could just follow the fire line out. That's easy. I learned that on the first day: fire line = dark and burned on one side, not so dark and unburned on the other. But with no pillar of fire to follow, the dark took on a new dimension in a remote marsh with nothing but impossibly distant stars to light the way. *Squish!* There was one of those wet spots the firelight helped me avoid on the way out. When everything is dark, burned and unburned look pretty much the same.

The flashlight, the one I left in the truck, would be helpful now, I thought, glad there was no one around to witness my lack of foresight.

It was then that it occurred to me that what I had done was, in some respects, the opposite of "painting oneself into a corner." In that corner, one knows exactly where they are and exactly where they need to get to. The problem is to do so without ruining the work accomplished. I had no danger of undoing the suppression work I had so expertly accomplished, but I really had no idea where I was and no idea of where I was going, except for a wistful picture of my truck in my mind. *Hmm,* I thought. *I might do this a little differently the next time.* Learning is such a beautiful thing.

With my best guess as to direction, I walked as straight as I could, soon with two wet feet, and began to see the silhouette of some trees, betraying higher ground. I finally reached the two-track and guessed my truck was to the left. I walked a long way, in the dark, in the silence, almost ready to turn around, thinking I'd guessed wrong, when I finally reached the truck.

About a week later, I had another night fire in a marsh. That time, I walked *into* the fire at the far end of the fire line and put it out on the way *back*. Fool me once.

THE MEASURE OF A RANGER

Mop-up is the last stage of fire suppression.[9] It involves putting out the last smoldering embers on a fire, which may sometimes be below the surface. A large fire may only involve a broad strip along the edge, the width of which is dependent on the nature of the fuels. It is dirty, unglamorous, and crucially important, and it greatly rewards thoroughness and attention to detail.

There is great elation in beating back flames higher than one's head and arresting the progress of the fire. Yet the truth about the beast is that in its near-dead phase after the great flames have been initially defeated, it is in its most deceitful and potentially dangerous state. One can easily lose focus and forget that the glowing fire persists, with all the latent potential to roar back to life. It is not until the last ember is defeated that the foe is ultimately vanquished. And the tremendous challenge is to find the last little "pocketlets" of fire and extinguish them.

The first step is to mark where the wisps of smoke are when they're visible. Sophisticated resource management agencies like the DNR have a special tool they use for this purpose—toilet paper. It's cheap, effective, and it breaks down quickly. Next comes the hard part. If you can reach the edge of the fire with a hose and water supply, that's a blessing. Otherwise, it's time for the shovel and back can. This is when most folks lose their fidelity, look for shortcuts, and then hope and pray. It's dirty work. The process is to dig out the hot spots and then wet them down. "Rinse and repeat until cool." The same spot, like a dead root burning underground, may need to be dug up and sprayed five or more times when it is first discovered, and it may need to be visited again and again. But the process needs to continue until all the fuels are cool. There's no glory. Oftentimes, mop-up is done alone, so you are your only accountability. When you're done, you will be filthy, smelly, and almost invariably bleeding. But, if you've done your work thoroughly, the fire will be out. That's the reward, and it has to be enough in itself to motivate the effort. For most people, it wouldn't be.

---

9 Mop-up is the last stage of "putting a fire to bed." It involves locating and putting out any hot spots on a fire, or along the edge of a large fire. It may involve digging out a stump or rolling over a large branch or a log to expose a still burning underside, previously protected from water spray, and cooling it. To be done well, it requires thoroughness and tenacity, a doggedness to find, expose, and put out every single smoldering fuel with the potential to later (an hour, a day, a week) rekindle an uncontained fire.

A mistake in completing mop-up well can result in a return trip to a fire, the ultimate humiliation for any ranger. Like proverbial yeast, one missed spot can escape and spread the fire all along the old fire line and beyond. That makes for a lot of additional work. Whoever is out helping on the escaped fire knows that they are out there because you didn't do your job the first time. When that involves people in the fire department who have left home and family or work to help, and with whom any ranger desperately wants credibility, that's a bad situation. Mop-up then needs to be done a second time, but now under the glare of culpability, a bitter pill.

Young rangers, understandably, appreciate the endorphins generated by knocking down the running fire, mano a mano. Mop-up is seen as an inglorious collateral effort, done mostly because it is "protocol." But if you are in the fire game long enough, every ranger experiences an escaped fire. More often than not, it is because of poorly executed mop-up. Young people didn't discount the importance of mop-up. They just failed to appreciate the perseverance needed, at least right away. There were a few rangers around who were known for their virtual infallibility on mop-ups. If you could find a wisp of smoke on one of their fires the next morning, it was extraordinary. Almost without exception, these rangers were older folks, people who had a few scars on their egos from past foibles. Most rangers learn the right way to do mop-up the hard way.

Part of the art of living is recognizing problems when they're still small, before they've significantly disrupted life. Like mop-up, there is a great benefit to aggressively addressing such problems while they're still small and doing so thoroughly until they have been fully dispatched. That takes a certain mental discipline, and as with mop-up, it takes some "seasoning" to appreciate its essential nature. The tuition may be high, but the School of Hard Knocks is an excellent teacher. Lesson 1 is that the shovel and bucket are effective tools. Lesson 2 might be to "phone a friend."

Marble Quarry Road

A fire on the north side of Marble Quarry Road in western Sauk County raised some interesting issues due to its challenging setting on a volatile day.

It later turned out to have been started by some young people four-wheeling. Their rigs got into some taller grass, which contacted hot exhaust elements on the underside of the four-wheelers and caught fire. The initial response was good, including two fire departments, me, and my heavy unit. But we quickly discovered that there was a branch of Bear Creek that flowed parallel to the road on the fire side. Its distance from Marble Quarry Road varied, and crossings on the driveways going back to the scattered houses along the road were of limited weight capacity. My 4 x 4 truck was (apparently) light enough to cross the bridge on the driveway that most directly accessed the fire. But the larger fire department trucks and my heavy unit were too heavy. So, while we got the smaller trucks to the fire right away, some energy had to be expended to identify another path to get larger resources to the fire.

When the fire department brush rig and I arrived, the fire had just reached an old barn structure of unknown use.[10] The south-facing wall was just getting involved and seemed like the most urgent exposure for us to address, especially since the other rigs would not be able to get to the fire for an unknown duration of time. Of course, as we focused on the barn, the wildfire continued to spread north, driven by a south wind. The local firefighters on the brush rig knew the area and indicated that there were no additional structures in the path of the fire as it raged up the draw to the north.

A fire department pumper arrived and took up the effort with the barn. That was my cue to take some action on the wildland fire. There was a two-track road going north through the bottom of the valley, and the fire had not yet crossed it to the right. I began working that edge in the hopes that the road would prevent it from burning to the east, and I could cut off its northern progress.

The fire reached an area in the valley where the grass became interspersed with red and white pine trees, six to ten feet tall.

---

10 "Brush rig" is a term for a smaller, more agile fire truck with a smaller water capacity but with four-wheel drive, enabling it to get "off-road" and gain closer access to a fire. They can vary in specific features, but most fire departments have at least one such truck. My smaller 4 × 4 truck would be considered a brush rig.

They were dry enough that flames quickly engulfed them, generating intense heat. I thought I was okay as long as I stayed on the road or to the east. But as I continued my efforts to "pinch off" the fire up the draw, I saw, to my dismay, that the fire had somehow crossed the road below me and was roaring up the east side of the valley, leaving me in the path of the advancing fire on both sides. Yikes! I could see ahead of me that there was some clearing, so I drove there and wetted down the lighter fuels around the truck. There was a rock hill going up to the top of the draw, which had no vegetation in it, so I climbed up on that, both to be safe and to observe the fire. As it roared up the draw, which it did with great ferocity, I was protected, and my truck, on its damp island, was also safe. Still, it wasn't the smartest move on my part.

The whole valley burned, and the fire was slowly making its way up the side slopes of the draw in both directions. Once the head fire reached the top of the hill to the north, it lost most of its energy. The other side, facing north, away from the sun, was not as dry. And the effect of the wind was blunted by the hill and the heavier vegetation that existed there. So, the fire progressed very slowly over the hill and was easy to stop. But the side hills in the draw caused a different problem.

By this time, access for the bigger trucks was established through a farm up the road. So additional units crossed there and traversed several fields to get involved in fire suppression. My heavy unit arrived, and my sidekick, Larry, unloaded the tractor plow to begin work on the right flank of the fire. It was burning in a rather steep area, in my estimation. Larry was pretty confident it was not too steep for him to operate. Trusting his judgment, I turned him loose. He made excellent progress, plowing a line up along that side of the fire, effectively stopping the fire's spread. But the farther he got, the farther up the hill the fire had burned and the steeper the side hill became. He was about to stop plowing because of the slope when, ever so gently, the tractor plow began to tip to the left downhill. Fortunately, there was a stout little oak tree there, which prevented the tractor from tipping more than a few inches off of its right roller pad. Larry was able to exit the tractor safely and began the humbling process of getting some heavy equipment to help disentangle the problem.

The fire ended up being about ninety acres but ran almost a mile to the north to the top of the hill. There was little lasting damage, except that the barn was scorched on the south side. But it brought to light a problem I had only considered with larger bodies of water, like rivers, which have commercial bridges crossing them. A little creek like this was only crossed via private driveways, most of which could handle only limited loads. It is difficult to anticipate all the potential issues one might have to resolve in wrestling with a forest fire. Or other life problems, for that matter.

How Does Your Garden Burn?

One of the geographic categories of fire protection in Wisconsin is termed "Cooperative." The cooperative areas in the state are those where the wildland fire danger is more limited. This is usually because most of the land is used for agriculture, limiting large contiguous areas through which a fire may carry. The DNR fire program does not have fire staff stationed in those areas but is available to provide support to fire departments, if needed. Otherwise, the responsibility for suppressing wildland fires there falls to the fire departments.

One late spring, there was a somewhat messy fire in the cooperative area east of Madison. It took place in a peat marsh. The running fire had been stopped, such that it was contained. But in the burned area, roughly fifty acres, smoldering fire had spread down into the underlying soil, a thick layer of peat. When you think peat, think peat moss. It is the same stuff derived from decaying plant material. While it has tremendous water retention characteristics when soaked, when it becomes very dry, it is almost waterproof. Such fires are notoriously difficult to put out completely and require uniquely specific, intense, and persistent efforts to suppress them completely.

Peat fires in Wisconsin are the stuff of legend. Like this one, they tend to be smoldering fires in the ground that persist after the running surface fire has passed. A story about one very large such fire has the initial attack ranger calling on the radio to numerous units responding to help: "Grab a piece of her, boys. There's plenty to go around." That remark is attributed to the beginning of efforts to establish a protocol, an organizational approach for fighting not just peat fires, but all wildland fires.

Another large marsh fire (several thousand acres) in the northwestern part of the state, several decades previous, occurred in the fall and was battled until the snow fell. The story is that there were sporadic columns of smoke and steam throughout the winter and that even the snowmelt did not completely extinguish it. Upon reengaging the suppression effort in spring, it was discovered that winter burning deep in the peat had created large underground cavities not visible from the surface. On one occasion, "they" say, an older, well-worn fire unit dropped into one such hole. The driver was rescued, but purportedly the decision was made to leave the vehicle, which was not visible from ground level.

Back to the present. In this dry spring, as the fire burned across the marsh, it burrowed into the bone-dry peat below, defying suppression. Lacking the equipment and workers over a protracted period to put out such a fire, the local volunteer fire department contacted the DNR for help. Since things had greened up at home, and wildland fire occurrences had mostly passed, at least in southern Wisconsin, we brought a team of eight people and various gear to help with the effort.

The equipment included two tractor plows whose blades could be used to dig up the peat where hotspots were located. That facilitated getting water onto those burning embers, otherwise protected below ground, and knocking them out. On two sides of the squarish marsh was a ditch, previously constructed to channel excess water away from adjacent agricultural fields. Our plan was to dam up both ends of the ditch, have a series of fire department tankers ferry water to it throughout the day, and use it as a water supply source. That would involve long and complex hose lays using pumps set up by the ditch to supply water throughout the fifty-acre marsh (a square forty-acre parcel is one-quarter-mile on each side).

A normal day in this effort involved a team of two people

going out at first light to locate the delicate wisps of smoke that betrayed a smoldering fire burning below the surface. Armed with copious amounts of toilet paper, they marked the location of all these wisps. At a more reasonable hour of the morning, the rest of the team would come, and the balance of the day was spent digging up those hot spots and putting them out. Initially, there were hundreds of such spots. The early morning marking team alternated among the group, a job that shrunk over the course of the week as we successfully doused more and more hot spots.

Over eight or nine days, the fire was eventually vanquished. Putting out peat is an extremely dirty job. Burned peat produces a very fine, almost powdery, whitish ash. Our activity stirred ash into the air, coating equipment, clothes, skin, and lungs. Evenings were marked, especially as the days dragged on, with a considerable chorus of coughing. We showered every night, but everything, including vehicles, tools, and clothing, was permeated with the smell of peat ash. In time, the smell of hot food was the only aroma strong enough to overcome it. At one point, a reporter from one of the Madison television stations came out with a cameraman to do a piece on the fire. Afterward, there were several comments, not coarse, that her perfume was the sweetest smelling thing we had experienced in too long of a time.

In the end, we calculated that we had something like two and a half miles of hose laid out in the marsh. On a couple of occasions, fire flared overnight, and a section of hose burned through, requiring new fittings to be spliced. The fire department gave us a tally of the gallonage they shuttled to the fire, one I can no longer remember, except that it was in the hundreds of thousands.

That was the only peat fire of any significance I ever worked, and that was okay with me.

FIRE DEPARTMENTS

Fire departments are all over the map, literally and figuratively.

The bigger, better-equipped departments are in the larger cities (my "big city" was Baraboo, with a population of about nine thousand). They have a strong tax base and more to protect. A couple of departments had enough work that the chief's job was a paid position. While I was in Spring Green, the first ladder truck in the area made its appearance in Baraboo. A four-story county office building had been built there, and several large warehouse structures justified the acquisition of such a pricey rig. In the first year, neighboring departments frequently requested the ladder truck for mutual aid assistance. As the novelty of its presence wore off and understanding grew about how to make use of it, its effectiveness grew. Within a few years, Reedsburg, twenty-five miles to the west, had a ladder truck, too. Working with these more robust departments was always easy. They were properly equipped, with everything fully operational, understood protocol, and responded in smart, clean, well-functioning fire gear. They behaved very professionally.

Apart from these larger towns, fire departments ran the gamut. Some were a little smaller and more humbly funded, with less equipment, but they were competently led and similarly professional in their response and performance.

On the other hand, there were a few tiny villages whose relative geographical isolation required a local response capability. These were less affluent communities, and their departments reflected that. Hearts were always in the right place, but the reality of their limitations showed. With a small population base from which to draw, they sometimes had difficulty maintaining a sufficient number on their roster of firefighters. There were a few members who should have retired long ago. Since there were often few jobs in these little towns, many who lived there worked elsewhere, and that could be problematic for response to daytime calls; getting half a dozen guys for an early afternoon grass fire on a weekday could be a challenge. Because of economics, these departments struggled to stay current with equipment requirements. Tankers might be old milk haulers with pumps cobbled on. Sometimes responders, especially older guys, would come to a scene straight from their farm homes wearing overalls. Without protective gear, they would serve away from direct fire duty by operating the pump on a truck, for example. But they all worked hard and were a valuable asset. They weren't too proud to call for help from a neighboring department when they needed it, which was frequently. Neighbors understood and were happy to help. Attitudes were right, and that's always workable.

The bigger departments had a good complement of brush rigs. Some were smaller and more nimble even than my truck. When I started, there were three or four Willys trucks, jeep-like military vehicles of 1950s vintage from the Korean War days (you can see them on M.A.S.H.). They were useful when they worked, but they had a different electrical system than most other vehicles, and there weren't many who understood how to repair them. But they were tough rigs. Occasionally, I asked a department to send just their brush rig(s) to a fire. They generally sent a "nurse" tanker, too, so they could independently keep the brush rig in water.

Most communities had a few people with some leadership skills who were willing to take a lead role with the local volunteer service. Sometimes there was only one, and they would serve in the chief's role for more than a generation, more than thirty, sometimes forty years.

There was federal grant money for fire departments that was funneled through the forest fire program, and rangers like me had some influence over where it went. My struggle was always between my heart and my head. The bigger departments would make better, more frequent use of the equipment, but I knew that if those departments really needed something, they could get it themselves. The smaller departments' needs were somewhat less acute, but these grants were one of the very few ways they could meet a need.

When a department asked a neighboring department for assistance on an incident, it was termed "mutual aid." And they were mutual. Complaints about being asked to help were rare, even by the "wealthy" departments, frequently tapped by more humble neighboring departments.

If a department had a "feed," a meal they put on to raise money for a truck or an addition to a fire station, you could be sure that not only would most of the local community come out for it, but all the neighboring fire departments and most of the rest in the county would send a healthy contingency. I always tried to attend those, too, getting a couple of additional carry-outs to take home. (With twenty-two departments in four counties, that could consume a lot of evenings.) There was great mutual support and camaraderie within the fire service.

Local politicians often attended these, too. While they may have had ulterior motives, their presence contributed to a sense of community.

Busted in Baraboo

Our willingness and desire to help neighboring rangers sometimes resulted in an unusual workload. I was in northern Sauk County when the Baraboo Fire Department was paged to a roadside fire east of town. That location was actually in my neighboring Poynette ranger's area, but he was tied up on another fire in eastern Columbia County, so I went to help.

It was late afternoon, and a small brush pile the local landowner had lit had escaped and burned through the grass along the north side of the road and up a steep, rocky slope to the north. The running fire was stopped at the top of the ridge where the woods met a field, but there was quite a bit of material burning on the hillside. Eventually, we were joined by another fire department, and the Poynette ranger broke free and came to the scene as well.

While there was not a great deal of exigency anymore, there was a lot of heavy mop-up to do, and it would be time-consuming. We were also running out of light. None of these conditions would have been much of an issue except for an accident that occurred midway up the slope.

One of the firefighters from Baraboo had slipped on one of the rock outcrops and had fallen hard several feet down the slope. There was a call for help for him, and since I was not far away, I headed in that direction. One of the firefighters, who was also an emergency medical technician (EMT), suspected that his lower leg was broken. The slope was steep enough that it would not be possible to walk him down the slope, aided or otherwise, so the decision was made to bring up a ladder. As one of the bigger and younger guys there, I offered to help. When the ladder arrived, the wounded fireman was strapped to it, using several of our belts, and we commenced to carry him down perhaps seventy-five yards of steep, rocky woods. We went slowly, but there was a certain amount of urgency, in part because our friend, the fireman, was hurting and in part because our light was diminishing quickly; darkness would make this difficult task almost impossible. I was on one beam of the ladder at the bottom. Another guy was on the other beam with two others at the top. We were doing fine until the other fellow on the lower end tripped and went down. Though he was not hurt, I suddenly had almost the full weight of this generously-sized fellow on the ladder on my shoulder. I managed not to drop him, but it felt like the beam of that ladder took a bite out of my shoulder. The fellow who fell got up quickly and returned to his post, and we finished our carry successfully.

It turned out that his leg *was* broken. I had a bruise and a sore shoulder for several weeks that served as a reminder of our adventure. It also turned out the landowner was a local politician who was notoriously and relentlessly critical of the DNR. This incident served to tone him down somewhat, and everyone was thankful for that.

With Honors

Sometimes you experience things that influence you and affect you in a way you weren't anticipating.

I was not more than five years into my job when I heard of the premature death of a man with the Ridgeway Fire Department. Ridgeway was on the southern edge of my protection area. In fact, a significant part of the area that the Ridgeway Fire Department was responsible for was outside of the DNR protection area (in the cooperative area). So, I didn't have a lot of interaction with them. I thought this funeral might be an opportunity to demonstrate the fact that I knew they were there and that I appreciated their help. I didn't know the deceased man, so it felt rather like I was just going down there to put in my time.

The service at the church was brief and otherwise unremarkable, as funerals go. But it was noted that he was a veteran of the Korean War and that they would be going out to the cemetery for a burial with military honors. I had heard of such ceremonies but had not witnessed one, so I thought I would invest in a few more minutes there, perhaps with the opportunity to speak directly to the chief or some of the other fire personnel, all of whom attended in dress fire regalia with the fire trucks and silent lights (emergency lights with no siren).

Outside the church, I got in my truck and waited for the procession to pass so I could follow. To my surprise, as the fire trucks slowly passed, one of the firefighters insistently waved me into the entourage, so I followed and turned my lights on as well. In my new position of honor, I joined them on the slow trip to the cemetery on top of the hill south of town.

The smartly dressed firefighters lined up to form a ten-foot-wide corridor from the roadway to the gravesite. Six men in military uniform, not from the fire department, took the casket and slowly, formally, deliberately, steps synchronized, carried the flag-draped coffin to the grave. Funerals, in general, and ceremonies like this, in particular, are moving. But in our increasingly noisy world, part of what is so gripping about them is the pervasive silence in which they take place.

After setting down the casket, they turned to face it. As the firefighters moved from their corridor positions and gathered near the gravesite, the only sound was that of the breeze blowing through the scattered trees at the cemetery and the poignant cry of a red-tailed hawk. Once everyone reached the grave, all stood motionless for a few moments, after which another soldier turned to his left and took two steps toward the gravesite. Tucked neatly under his arm was a bugle, which he then brought forth. He proceeded to play taps, each note lingering a painfully long time. "Day is done, God is nigh." The closing note was held for so long that it was difficult to tell when it ended, as the sound drifted away in the sunny afternoon breeze. Following a silent salute, he returned the trumpet to its position under his arm and painstakingly returned to his previous position, all in deafening silence.

I had heard taps before, to be sure, but I had never experienced anything like this. I was frozen where I stood, as everyone seemed to be. Tears found their way down my cheeks.

The two soldiers on the left side of the casket lifted the flag, the one at the feet beginning the deliberate folding process, thirteen times, resulting in a neatly folded blue triangle. The soldier at the head moved in square steps to the widow, leaned over, and presented it to her with a handful of quiet words. "On behalf of the president of the United States, the United States Navy, and a grateful nation, please accept this flag as a symbol of our appreciation for your loved one's honorable and faithful service." The whole process sucked the air out of my lungs. We all stood motionless, silent in the warm sunshine and gentle breezes, reflecting on the sacrifice that had just been so eloquently honored.

The pastor said a few words, projecting his voice so all could hear. He finished with a psalm and closed with a prayer. Eventually, people started to mill away, all in silence, with only an occasional grim nod of acknowledgment between people. I never said a word to anyone.

One does not have to think long about the purpose of the military to realize what it all too frequently means. The accounts of war are horrifying, and they don't do justice to the experiences of those there. Such dreadful activities should never be glorified. However, it is crucial to honor the commitment and faith that leads one to give all they have for something they believe in. In this case, it was the risk of life for one's country. But it is done every day for relationships, too, and often for principle. There is something inherently valuable, nourishing to the soul, to have something recognized in one's life that is so valuable, so crucial to meaningful life, that one is willing to sacrifice everything for it.

Twisting in the Wind

"There's something happenin' here. What it is ain't exactly clear."
—Buffalo Springfield, "For What It's Worth"

As a new, young ranger, efficiency was one of my objectives. Efficiency is a good goal, but when an excess of it is combined with inexperience, there can be problems. On an early spring day, I was working with a crew of mostly wildlife people to conduct several controlled burns in grasslands. With a deep desire to get them done as early in the day as possible, as afternoon weather had a good chance of bringing wildfires, I decided to start us on our burning activities earlier than the crew was used to. I was encouraged to wait another half hour or forty-five minutes to initiate the first burn. Oh, but I knew better.

The control lines were already in around the first piece we were to burn, a square-shaped field of prairie grass of roughly four acres. We began burning around the edges to widen the breaks, beginning with the narrowest one. The air was almost completely still, so there were minimal concerns about the fire getting away. All was proceeding splendidly, and the fire was gradually burning in from all sides toward the center of the prairie, a safe, slow, and complete burn.

But as the fire approached the last half acre or so, I noticed that the flames on my side of the fire were leaning just a little to the right, and the flames on the far side of the fire were leaning just a little bit to the left. *What a peculiar phenomenon*, I thought. As the edges of the fire burned closer together, I saw the "slanted flame" phenomenon intensify. Soon, a significant and unified column of smoke and heat began to gather over the center of the fire, and the flames around the edges of the fire were drawn toward the center, consuming fresh fuel at an increasing rate. In a moment, as I watched, the flames originating on the ground met a few feet above the grass and formed a column of fire spinning in a counterclockwise direction. In seconds, as more grass and more heat rapidly became involved, that column of flame, perhaps eighteen inches across, climbed to a height of sixty feet or more. A soft, high-pitched whistling sound accompanied the twirling inferno. I had never seen anything like this, and I was dumbfounded. But I "came to" and realized the significant threat this storm had for carrying a burning ember outside, perhaps *far* outside, of our control lines.

As I assessed the crew's readiness, they were similarly absorbed, many with their mouths agape, completely absent from their responsibility to watch the fire line. But before I could respond, the flames had met, consumed the last of the fuel, and the fire and the whistling column were a quiet memory. It was all over. Frantically scanning the surrounding area, I noticed that there was no smoke to be observed immediately. I dispersed the crew to check around the edges to be sure. As they searched, dust devils comprised of ash and hot air twirled and danced across the warm, dark, burned area.

Later, one of the more seasoned members of the crew discreetly mentioned to me that they usually wait to start burning until a little later because it is then that the morning's gentle breezes begin and serve to subtly break up the development of columns of heat that resulted in our firestorm. Another lesson learned, this one in physics, and an inexpensive one at that; I was now better prepared.

Later in my career, I read stories of similar behavior in large wildfires out West. Those accounts, however, were exponentially more extreme. In one case, a column was characterized as being eight feet across. It lifted burning pieces of wood weighing a hundred pounds more than a hundred feet in the air and slung them out like matchsticks, igniting new areas. As scary as it was to me, my little whirlwind was relatively tame.

It's curious how different an event can appear when viewed retrospectively. Patterns can be seen afterward that aren't apparent when navigating them going forward in real time. Events in life can be the same way. It's naive to suggest that all bad experiences have good results. Experience teaches us otherwise. But when an event is over, and matters are settled, and we assess the new reality, it's unusual for there not to be something to be learned to better traverse the days ahead.

White Lightning in Leland

A little country music in the background, the rumble of voices, and the occasional clink meant one thing: somebody was calling from the tavern. The noise made it hard to hear.

"Hello?" I said.

"Hello?"

"Hello? Can I help you?"

"Is this the DNR?"

"This is Blair Anderson, the DNR ranger. Can I help you?"

Pause. "So, this is the DNR?"

There was no point in trying to fine-tune the information. This conversation was not going to take place at that level. Think Otis from *The Andy Griffith Show*. It was Saturday afternoon.

"Yup!"

"Well, this is Harold." Then, to someone else, "Yeah, I got him."

"Hello, Harold. What can I do for you? Are we settling a bet?" Occasionally, I would get a call from a bar to adjudicate a lubricated and inflamed dispute about a game regulation or an animal characteristic or some such matter. The wardens got them more often than I did.

"No, no, nothing like that." Having addressed that, he took a moment to gather why he had called. "We got a fire."

"Oh," I responded with just a hair of increased interest. I'm picturing a burger on the grill that got a little out of hand. "Where are you?"

"Oh, we're up at Junior's." Okay. They're at Sprecher's in Leland. I knew the place, and I knew Junior because he was one of my fire wardens. His and another bar were the only public establishments in Leland.

"You've got a fire there, huh?"

"Yes, we do."

"Did you call the fire department?"

"No. We thought it would be better to call you."

"Okay." Let's hear how the boys at the bar arrived at this conclusion. "Tell me about the fire."

"Well, we can see it from here." Good start.

"Okay. Where is it?"

"Well, it's in a tree across the pond." A long, wide stretch in a branch of Honey Creek bordered the west side of Leland. To the west of the "pond" was a large, somewhat swampy area populated with stunted trees. I was hoping this guy was mistaken, which seemed like a pretty good bet, because the only way to access that area is via a long walk from the opposite side, maybe a half mile off of Highway PF.

"Yeah, it's in the top of a tree over there." Terrific. There had been a thunderstorm on Thursday night, and some lightning strikes could have occurred, with this fire the result of one of them. If this guy was right, and I still wasn't sure, I'd have to hike a half mile with a chainsaw and some other tools to knock the tree down and put out the fire. There was rain in the forecast. But it was dry enough then that if a branch broke off and hit the ground, it could start spreading.

"Okay. Hey, let me talk to Junior for a minute, will ya'?"

"Oh, yeah. He's right here. Hang on a minute."

After some background discussion and other assorted noise, "Hey, Blair. I hope it's okay we called you at home. I would've called, but Harold was pretty insistent about reporting it himself. You know how he gets." I didn't, but that was okay.

"No problem, Junior. So, what's the deal? Is this guy right about the tree on fire?"

"I think so. I took a quick look, and there's smoke in the top of a tree over there, and it's been there for twenty minutes at least."

"Okay. Thanks. I'll be over that way in a little bit." It was one of those partly sunny, partly cloudy days, warm and humid, so it didn't seem like much of a fire day. But you don't want to ignore a report that turns out later to be a bona fide fire. Since it was Saturday, we were getting a rare day off in spring ... at least some of us were. The heavy unit would be of no use in the swamp, but I did not want to do this alone. I toyed with the idea of having the Plain Fire Department meet me there. But considering the circumstances of the report, I decided it was best to have a look for myself before interrupting a bunch of folks' weekend activities.

I drove to Leland, and somewhat to my surprise, it was just as reported, except that by then, a chunk had burned enough that it broke off, and there was fire tentatively walking around on the ground underneath the tree. That sealed the deal for the good ol' boys from Plain, and I had Sauk County page them out.

I met them on Highway PF a little bit after one o'clock and decided to have a small squad hike in to cut down the tree and extinguish the little fire. One of the firefighters knew where there was a makeshift bridge we could use to cross the little creek between the adjacent field and the marsh in question. Six of us jumped on their brush rig and drove around the edge of the tilled field to the bridge. We walked in from there with a chainsaw, a couple of back cans, and some assorted hand tools.

After ten minutes of picking our way through the trees and puddles, the fire was in sight of us, when I sensed a change in the air. Looking back to the west, I saw an ominous cloud beginning to appear over the hill. I had a difficult, safety-related decision to make. A storm cell like that could make for some dangerous

A photo from the tavern across Leland Pond toward the location of the fire.

conditions. It was still my fire, so I told the guys we were going to turn around and head out. That was not well received, especially since we were only a few dozen steps from the fire. But I stuck to it, and we retreated west amidst some soft muttering.

By the time we reached the bridge and the brush rig, attitudes had changed. The temperature had dropped fifteen degrees, and there was a damp chill in the air. Partly sunny had turned to completely cloudy, and then some. Even if this was just a small cell, it was going to be a wild one. Equipment was thrown rather haphazardly onto the brush rig, everyone piled on, and we took off for the trucks parked on the highway. By the time we arrived, the first few drops were beginning to fall. Within thirty seconds, equipment had been put away, everyone was in their respective vehicles, and we were in a deluge. As we headed to our respective stations, it was so dark that it could easily have been nine o'clock at night.

I knew our ground fire didn't stand a chance. I went back later that day, alone, after the storm had passed, and again the next day to make sure the fire in the tree was also out. After the last visit, on Sunday afternoon, I stopped by Junior's and thanked

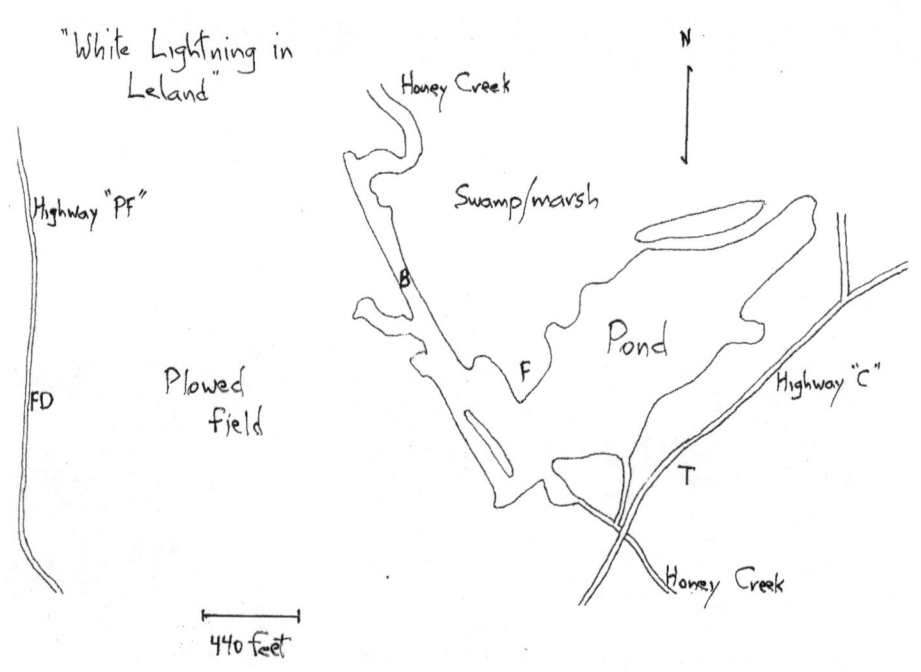

A sketch of the area around the fire. "T" is the tavern from which the fire was first spotted. "F" is the location of the fire. "FD" is where the fire department staged on the road to the west. "B" is the makeshift bridge we used to cross into the swamp to (almost) access the fire.

everyone for their quality assistance—especially Harold, who beamed and toasted me with a drink.

Tracers

Most of the twenty or so fire department chiefs I worked with were cooperative and were willing to concede my statutory authority on wildland fires, especially when we worked respectfully with each other (a wise idea in any situation). I had one guy, though, I'll call him Oscar, who didn't think like that. He was a pretty good guy and not a bad chief. However, no one had a higher opinion of Oscar than Oscar. His fire department folks seemed to like him well enough. I don't think anyone envied the chief's job, especially in this larger town, and he led with considerable authority. Most people are comfortable following strong, confident leadership. And few were more confident than Oscar.

He ran a very successful independent business in town and seemed to be quite wealthy. He flashed cash regularly and liked to talk about all the trips he took, including an annual pilgrimage to a top sports championship. Most country people, including those who live "in town," don't have those sorts of resources, and the few who do talk about it as much as Oscar did.

I tended to let Oscar wear the crown at wildland fire incidents as long as he was doing the right thing, tactically, which he usually did. Occasionally, he'd be in over his head, though he didn't know it, and I would have to suggest a different approach. That never went over well, and Oscar's folks, for self-preservation purposes, tended to follow his direction. On one occasion, I called for mutual aid from another fire department because I needed additional manpower to do what I wanted to do on the fire. That was not well received. Oscar and I talked about it later, without resolution. Neighboring fire departments knew how he was. And one of his own officers apologized to me, on the side, for what had happened. It takes all kinds, and we got the fire out.

But there was an incident one year that pretty well resolved the authority problem. We had had a very long, exceptionally dry stretch in spring. On this particular evening, even the usual rise in humidity associated with dusk failed to materialize. The winds continued to blow very strongly, close to thirty miles per hour from the west, and the humidity hovered in the teens, even after dark—a most unusual occurrence. These were alarming conditions, especially for that time of day when it was hard to see. There was no talk from our supervisor about what time stations would close that evening because everyone was working active fires through the late afternoon and into the evening.

Around ten o'clock that night, Oscar's department was paged to a grass fire east of Reedsburg. I had just finished with a fire near Spring Green, twenty-five miles away, and headed in that direction. The Richland Center ranger, a newer employee, was free, so I asked him to respond as well. He could get there at least as soon as I could, and I was concerned about a large, wooded tract with few roads downwind from the fire location. When I arrived, I witnessed a fire phenomenon I'd never seen before. The fire, a little to the north of where we could gain access with our trucks, was burning through a marsh. But I could see that something in the marsh, perhaps the heads on cattails, were igniting and taking flight in a straight line horizontally across the marsh in the high winds, like tracers. Later, when the fire reached the uplands, these tracers would rise with the wind when they reached the contour of the land and continue on until they burned out. It was the only fire I was ever on when eye protection was an absolute necessity.

On my arrival, I located Oscar, who had been there for twenty minutes, to get a briefing and plan strategy. I'll never forget what I saw. Oscar clearly had never seen a fire like this burning under these conditions. It must have appeared absolutely unstoppable to him, and he had no idea what to do. As I asked a series of simple questions about the fire's status, his drawling response was, "Well," and he'd look off in the distance, or "I don't know," or "I'm not sure." This was extreme fire behavior for me, too, but it was very un-Oscar-like behavior.

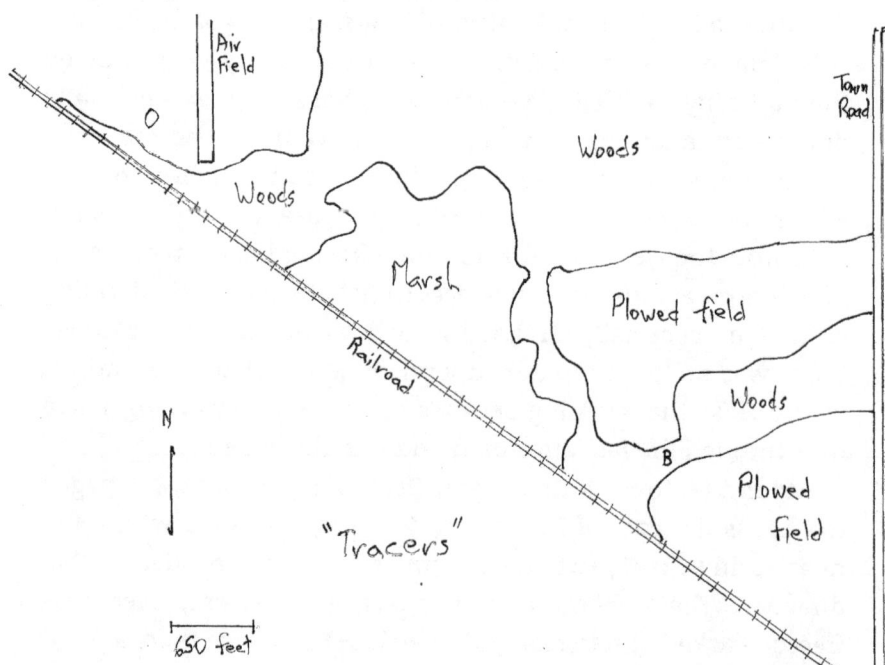

A sketch of the area around the fire. "O" is the general area where the fire originated. The woods and the marsh were dry and would burn. "B" is the narrow point in the woods where we set our backfire and were able to stop the main fire's progress. A railroad runs to the southwest of the fire. Because the wind was blowing nearly parallel to the tracks, they served as an effective fire break.

I connected with one of his assistants, whose expression said, "Uh-oh. What do we do now?" I started off toward the fire, motioning for the assistant to come with me, which he did.

We looked at the approaching fire in the distance and where it was likely to go as it burned toward us. Once it burned through the marsh, it would hit woods about one hundred yards before it reached us. The woods was more than ready to burn, and behind us, ahead of the fire, past the woods, was an ocean of grass. We could never stop it there. And beyond that were the Baraboo Bluffs, a vast and remote wooded area with very few roads. If it reached there, it could run for ten miles. And there were a number of scattered houses that would then be in the fire's ravenous path, and they'd be hard to protect in the dark.

But there was one opportunity. Not far from where we were, the woods narrowed to a width of about fifty yards with cultivated fields, now bare soil, on either side, excellent fire breaks. If we made a stand there, we had a chance. We figured we had fifteen, maybe twenty minutes before the fire reached there.

I asked the assistant to get his folks working on grubbing in a crude line through the woods from one field to the other as fast as they could work. He was a smart guy. "That'll never hold in these conditions."

I grinned at him. "Nope. We're going to backfire from it." The confidence drained from his face as he took in what I was suggesting.

"We're going to start another fire?" he gasped, incredulous.

I looked toward the tracers rapidly approaching us and turned back to him. "What do we have to lose?"

I was impressed with the speed at which he took it all in, including the likely futility of our effort. He shrugged, and in an instant, he charged off to brief his folks, and they went to work. Everyone understood the urgency of the situation.

My neighboring ranger had arrived and found Oscar in the same state. This young fellow was less experienced, so he was more than willing to follow my lead. I took him to the far side of what would be our "break" and looked back from where we had come. "We're going to backfire from here."

He looked at me, looked back at the break, and looked at me again. "Okay," he said in an if-you-say-so manner. What he lacked in confidence, he more than made up for in willingness, and he looked at me for the further direction he desperately hoped was coming.

"Once these guys get this line scratched in," and they had already made lots of progress, "we'll start firing from the middle in both directions. You get a drip torch and take a crew with back cans back toward the trucks, and I'll do the same toward the field here. String 'em out as you go to make sure it holds initially. Got it?"

Again, "Okay," but this time, I could see that he was translating what he'd been trained about, in principle, to the real flames-and-heat-and-oh-my-gosh-there's-a-fire-coming mentality he would need to help pull this off.

As our running fire emerged from the marsh and marched through the woods toward us, we braced. Stiff winds continued to blow material out of the marsh. Tracers were so thick that it was difficult to discipline oneself to continue to face the fire. As it neared, the advancing fire, super-heated air pressing skyward, searched frantically for more oxygen and began to draw the backfire toward it in defiance of the prevailing wind. The flames on the backfire leaned toward the approaching fire, just like they were supposed to, and more rapidly consumed more of the intervening fuel between the break and the main fire. Less than a minute later, in a crescendo of heat and noise, the flames from the two fires embraced violently in a turbulent climax, flames spinning upward, a deafening *WHOOOOSH!!!*

And then it was done. A vast cloud of smoke raced east from us, invisible in the night sky. Quickly, it grew quiet, and the cool of the evening reasserted itself. Suddenly dark, we adjusted our focus and were easily able to spy and extinguish any escaped spots across the break. Within a minute, it was clear we had stopped the advancing fire, and there was a hesitant sense of victory, albeit significantly mitigated by energy letdown and the sudden acknowledgment of fatigue.

I found my friend, the assistant chief, and we made some plans to do a little mop-up and patrol and then to set a watch in place for the night. We worked it out well together, and I was thankful. I didn't see Oscar again that night.

The next morning, I went to the fire station to get some information before heading out to the scene of the fire again to investigate the cause. Who should I find there but Oscar.

"How's Oscar this morning?" I greeted him jovially.

He took a moment before he looked up from what he was doing at his desk. The swagger was back, and I was a little relieved to see it, much to my surprise. "Oh, pretty good," he responded dismissively. He had to be tired. We all were. It had been well after midnight before we cleared the fire scene, leaving two of Oscar's guys to watch it through the night, and it wasn't yet seven in the morning.

"Any word from your guys out there last night?"

"Oh, no. We had it last night," he said with casual confidence. I loved the "we." Turning back to what he was doing, he said, "I didn't think that was going to work last night," referring to the backfire. That was as close as I was going to get to a "nice job."

After a lull, I said, "Well, I'm going to go take another look in the daylight." I thought he might want to see it, too, though I wasn't sure he'd want to do so with me.

Again, without looking up, he said, "Okay." That was it.

But he never gave me a hard time on a fire again.

Squeeze Play

One of the stranger assignments I had was to help with problems at Devil's Lake State Park after a torrential downpour one night in 1993. There is only a narrow passage with a normally small stream to drain the whole Devil's Lake watershed, which isn't all that large. But when twelve inches of rain falls in eighteen hours, and it all has to squeeze through that tiny creek, bad things can happen. That's what happened one summer night.

And what a mess it made. The little stream swelled so far out of its bed that it undercut the railroad tracks that pass along the east side of the lake and next to the creek, causing them to sag down into the creek bed. By a tavern just outside the park, cars that had been parked in the lot were stacked up against the building three high by the raging waters. No one was hurt, but much camping equipment was abandoned and washed away in the race to get sleepy campers to safety.

The next day, there were numerous residual problems there. The roads into and out of several of the campgrounds had a coating of extremely slick mud and silt that had been left there by the receding flood waters. Several cars tried to navigate the roads and slipped off as though they were on wet ice. With our

The railroad tracks adjacent to the drainage at Devil's Lake State Park after the heavy rains and severe flooding in 1993.

large water capacity trucks and powerful pumps, we could rinse the mud off the roads, making them passable and enabling the rescue effort for stranded campers and equipment.

Fire season was over, so we weren't thinking about anything related to our fire equipment. But the tanker was ready, and we headed up to the park early in the day. The drill involved filling the truck to its eight hundred-gallon capacity and working progressively along a road, spraying away the mud until the tank was empty. After that, we could simply return to a parking lot in the park where the water was still over a foot deep. The watershed was continuing to drain, and the standing water in the parking lot was contiguous with 374-acre Devil's Lake. Our truck had high clearance, so driving in the standing water of the parking lot was not a problem, and I was able to simply flop over a short section of draft hose to refill the trucks. The process repeated itself endlessly throughout the day.

The press was interested in what had happened at the park, and this allowed me the opportunity to have some mildly inappropriate fun. I was setting up to fill the truck again, for the umpteenth time, out of the standing water in the vast parking lot, a process which took six or eight minutes, when I spotted what I presumed was a reporter on an adjacent high road. I started the drafting process, which could continue on its own, and strolled over to the reporter. He was looking curiously at my truck and asked what I was doing. It was late in the afternoon, I was tired, and it seemed to me like a stupid question. I couldn't resist.

"Because of the flood, there is some heavy equipment scheduled to come in tonight to help clear away debris in the park. We need a place to stage the equipment, so I am removing the water from the parking lot so we have a place to put it. That truck holds eight hundred gallons, you know. I take it from here and dump it outside the park." His eyebrow raised. I looked back at the truck. "I gotta go." With that, I left him. When I finished drafting, I looked up to see him still staring after me. I never figured out who he was or if he genuinely thought me a fool. But it was worth the flummoxed look on his face.

A Hot Fire's Range

I had been fortunate to have never been involved in an especially destructive, multiple structure fire under extremely dry fire conditions. But that was about to change.

Looking at a burning permit request in the far north part of my area, I noticed a small column of smoke to the north of me, out of my area (Dodgeville). It appeared to be coming from a place close to where I knew the village of Lyndon Station to be. I called Louie on the radio and asked if he had heard anything about a fire there. Lyndon Station is in a different supervisory area (Wisconsin Rapids) with a different dispatcher, so it would not necessarily have been something Louie would have heard about. He called Lloyd Dettwiler, his counterpart to the north, but no one there knew anything about a fire near Lyndon Station. Rangers were not to leave their area without specific direction, so I staged on the side of a road, on top of a hill, at the edge of my area, and watched as the smoke column rapidly grew.

As it became the most compelling element of the sky to the north, it was clear to me that this was a significant fire. After another call that generated no information, I announced to Louie that I was heading in that direction, without orders. Almost the moment I did, the Lynden Station Fire Department was paged out to the fire, and things started happening quickly.

I arrived soon after the local ranger, Bill Hoffmann, and was given the task of trying to keep the rapidly advancing fire from crossing Highway 16. The winds that day would cause the fire to hit the highway at a soft angle, and with the four fire department trucks that were assigned to me, I thought that making that stop would not be difficult.

Still far ahead of the advancing fire, I got together with the fire department units at the place on Highway 16 where I thought the fire would hit. We would work together, using a little water to hose down the fuels across the road from the advancing fire and save the rest to put out any spot fires that may cross the highway. I felt quite sure that we would be able to stop it, and I told them so, both because I believed it and to give them (and me) some confidence. Timing would be critical, and I told everyone to wait until the fire was close before starting to spray water. Apply it too soon, and it would evaporate before the fire's arrival and be of no benefit. The head of the fire was still several minutes away, not yet in sight, but there were a few smaller spots that ignited and then grew between us and the main fire.

We were primed to meet the fire, and I thought it would still be a minute or more before I wanted them to start spraying preventative water. But one of the younger firefighters alertly observed a spot fire starting *across* Highway 16, and their unit went immediately to stop it, which they did successfully. I was very surprised that there was already a spot fire when we could barely see the advancing flame front. Within moments, however, two more spot fires were observed, and quickly, several more, some a considerable distance beyond us, away from the edge of the road. We had scarcely begun our suppression action, and it was clear to me that we had already lost Highway 16. Within thirty seconds, several of the spot fires had burned together and developed an independent new fire front. I was stunned at the efficiency with which the fire had passed us.

Ahead of this new flame front were several houses in a row. Quickly, and without permission, I reassigned my fire department units to protect those houses as best they could. I made a quick radio call to Hoffmann, who was now in the Fire Boss position (Incident Commander in modern terminology) of the freshly established command structure of this major fire, and gave him my discouraging update. The fire departments and I focused on the first two houses and were effective in keeping them from becoming involved as the leaves and grass burned away around them. I dispatched them to the last two houses in succession, and I approached the third house, where some leaves under an adjacent deck had ignited. They were burning well and were just beginning to involve the boards on the deck. Some of the house's vinyl siding was beginning to sag from the heat. A little water sprayed under the deck was sufficient to put that out, likely saving the house from full involvement.

Having saved the five houses, I reassessed my "group's" role. Although effective, we were freelancing, and that's not helpful in the overall management of the fire. The fire had burned through wildland fuels far past our location, leaving us in the wrong place for our assignment: to stop the advancing head of the fire. So, I contacted the fire boss, briefed him, and asked him what he wanted me to do.

Over the rest of the afternoon and into the evening, many more units, much equipment, and many fire departments joined the effort. I was eventually released from the fire at two o'clock in the morning. By then, the fire had been stopped, though there were still many hot spots needing urgent action so they wouldn't come back to life in the dry breezes the next day would bring. I was thankful that that particular headache, the next day, would not be mine.

Fire behavior like what I witnessed in Lyndon Station brought me a new appreciation for some of the stories I had heard from the past and some of the previously inexplicable (to me, anyway) ineffectiveness of fire resources that were part of those experiences.

It is not wise to judge another's efforts from such a place of inexperience and ignorance. My mild affinity for critiquing others dropped several notches after that day. I was better for it.

He Can't Be!

"Until I heard the sudden word that a friend of mine was dead."
—Jackson Browne, "Song for Adam"

I was in my second fire season, and I was beginning to see that this could be a fun job. Fire suppression, which originally seemed intimidating, and even a bit neanderthal, had become a swashbuckling adventure that required a set of skills and some mettle. It was sort of dangerous, enough to be exciting, but not so much as to be scary. I knew people had been killed on big fires out West, but that was ... out West. This was Wisconsin.

The new facility out at Tower Hill State Park had been completed the previous fall, and we were now moved in. There had been some conveniences associated with being in town, but access to the garage was difficult, especially for the heavy unit. Out here, we had lots of space and some privacy. In town, people would wander in with nothing better to do, just to talk about a deer they had spotted or some new regulation. During fire season, the park was not yet open to the public, so the park area was a wonderful place to go to eat lunch, take a walk, watch spring unfold, or just reflect a little.

This day was unexpectedly quiet, with only a couple of small fires on a day when more fire activity might have been expected, along with the potential for challenging suppression action. The woods was well dried out, and green-up had not started yet.

It was almost five thirty, and we were planning to close the stations when the phone rang. It was my boss. After some brief small talk about how the day went, he began to tell me about a fire that had taken place toward the northwest part of the state. It had grown to an estimated eighty acres (it turned out to be ninety-two) and burned in grass and into short pine near Augusta. As he got into increasing detail about a fire that wasn't close to our area and which was already suppressed, I began to slip into a "this affects me how" mindset.

My boss was being gentle, especially with this new guy, but finally got to the point.

"One of our firefighters was flanked by the fire, it overtook him, and he died."

One of those brief, seemingly interminable silences followed. I was staggered. There's a dramatic technique occasionally used in movies when the camera very quickly goes from a broad landscape to a close-up of a person's face or a crucial object. That's what happened in my mind. It had been slowly drifting away from what seemed like a nebulous conversation to wherever else the mind tends to wander. And it suddenly, forcefully, snapped into sharp focus.

Dead? One of our fire guys is actually dead because the fire killed him? In Wisconsin?

Finally, I asked, "What happened?" My boss didn't know all that much yet. It was one of our equipment operators, and it sounded like he left his equipment before he was flanked by the fire. His name was Don Eisberner.

Without many additional details, our conversation soon ended. He had similar calls to make to others. I was left with my thoughts. For a while, the starkest reality crowded and prevented any other thoughts: this guy was dead. This guy was

BLAIR W. ANDERSON

Several photographs from the Don Eisberner memorial in Eau Claire County related to the Canoe Landing Fire of April 24, 1982. The information sign discusses what happened with the fire and how Don became trapped. The enlargement of the map shows not only the 1982 fire and where the problem occurred but also another larger fire that burned in the same area twelve years later. Don's was thankfully the only line-of-duty death of any department fire personnel until one of our pilots crashed on a fire in 2009.

WHEN THE SMOKE CLEARS

dead because of a fire, a wildfire. Don Eisberner. I didn't know him. I was new enough that I didn't even recognize his name. The only picture my mind could generate was that of Larry, the equipment operator I worked with every day at Tower Hill. Larry and I were friends, and the irresistible urge to attach his face to a scenario that was lacking in detail made it even more difficult to contemplate.

It had been decades since a Wisconsin wildland firefighter had died working a fire.

How could this happen? As I sat in my chair, my mind drifted back, thinking about what we had been taught in our training at Tomahawk. Some of the themes and lessons were repeated so frequently that it almost got to be a joke for our group of initiates. I had heard them repeatedly at Tomahawk and again and again at other training sessions. But they were like rules to a game. Yes, it's important to follow them, but if you forget one, you lose the game, you learn a lesson, and you do better the next time. But now Don Eisberner was dead because somebody, maybe somebody else, maybe him, didn't follow one of those rules, did something they shouldn't have done, or didn't do something they *should* have done.

Then, as it came to me that I had to get it together and tell Larry, he breezed by my door. "See you tomorrow!"

"Hang on, Larry. Come in and sit down for a minute. I have some news." Noting my somber tone, he responded in kind as he sat down. Larry was a veteran of many more fire seasons than I was, and he was a few years older than me. Out of respect, I had navigated our relationship carefully. It struck me that he probably knew Don personally.

"What's up?"

"There was a fire out of Fairchild today, and one of the equipment operators was overrun on the fire and died."

"Oh, no!" he cringed in sudden, genuine anguish. "Who was it?"

"I don't know him, and I hadn't heard—"

"Don Eisberner?"

"Yes, that's the name." I immediately felt insensitive for referring to the deceased as "him" rather than using his name.

"Oh, no. I know Don," came the painful response. And then, "I knew Don," as the reality struck him, too. After a pause of internal acknowledgment, Larry asked, "What happened?"

"I don't know. Information is still sketchy." I acknowledged Larry's loss, which was, in a personal sense, greater than mine. "I'm sorry."

Larry spoke briefly of the last time he had seen Don, and we sat in silence, each trying to figure out how this fit into our understanding of our jobs, into our lives. What we both had to do was mostly internal to our own thinking, so after a few mostly quiet minutes, he left. We ventured into our respective evenings, different than either of us could have anticipated.

In the days that followed, we were busy with fires, which was probably a good thing. But there was at once both a flatness and an edge to the voices that came across the radio, as people were both discouraged and anxious in reaction to Don's death. The approach to the fire day had altered.

As with most pieces of heavy news like this, the initial emotional reaction dissipated fairly quickly. But how I looked at my work was forever changed. I knew intellectually that bad things like this could happen and could, in theory, happen in Wisconsin. But this was now, here, with a piece of equipment like the one in the garage at Tower Hill, with a guy in the same position as my friend Larry. A shadow of seriousness had been cast over the excitement of my work, and I never looked at it quite the same way again.

Many years later, when I was the chief ranger, one of our pilots crashed and was killed working a fire. I was not a pilot, and it did not seem as directly connected to the fire as this incident. Don Eisberner died at the Canoe Landing Fire because the fire burned around the break he had plowed in, and when he found himself flanked and about to be overtaken by the fire, he abandoned his equipment and was not able to outrun the fire. But another key detail emerged from the investigation.

For some reason, Don was not driving the tractor plow unit he usually did. And the mechanism to control the plow in this unit was configured in the opposite way from the one he was used to. That is, in his usual unit, pulling back on the plow control lowered it into the ground, and pushing the control forward raised it. But in this unit, pulling back raised the plow. In the relative calm of building a line along a fire, that wasn't a problem. But when he saw he was flanked and about to be overtaken by the fire, Don went into instinct mode, fight-or-flight. His natural reaction to raise the plow so he could turn, speed up, and get out of the path of the fire was to push the plow lever. In this unit, the move sunk the plow even deeper. Suddenly stuck, for a reason his panicked mind didn't instantly grasp, he abandoned ship. And that cost him his life.

Research subsequently showed that the likelihood of survival from a burn over is greatly increased by staying with the equipment. As a result of Don's death, all the controls on all the units, for both the plow and the blade on the front, were configured identically. Additionally, a safety system was developed for the tractor plows, which involved some additional plumbing. Around the top of the driver's reinforced compartment, hose was attached all around the top of the roll cage, with six or eight nozzles aimed at and around the operator's seat. It was connected to the pump and could be turned on with the pull of a lever and the quarter turn of a valve, both easily within the operator's reach. It was affectionately called the emergency immersion system, EIS (as in EISberner), for short. It was a standard part of every tractor plow after that, and part of the morning testing routine of equipment involved briefly activating it to ensure it worked. That generally involved the operators getting at least a little wet, but a different climate had settled over our work, and no one complained.

I had previously felt a certain sense of gamesmanship in my interaction with people. But that sporting sense was largely gone with this stark realization of how significant the consequences of those irresponsible actions could be. I approached people a little less jauntily after that.

COMMUNITY

HOW IT WORKS HERE

"I am memory and stillness. I am lonely in old age. I am not your destination. I am clinging to my ways. I am a town."
—Mary Chapin Carpenter, "I Am a Town"

I knew Spring Green would be a far cry from Chicago, where I grew up, and Madison, where I had been for the intervening three years. (Madison is different from anywhere.) My mother grew up in rural Iowa, and we had visited some of her family that still lived in small towns there, so I had some appreciation of what might be before me, like seeing more people I knew in the course of a day than I was used to. What I discovered was that seeing fewer people I didn't know would be so different. But it was the *nature* of the relationships that I couldn't have anticipated.

After moving to Spring Green, I quickly realized I had no frame of reference for what day-to-day life was like in such a small community. In fact, I would discover a whole new appreciation for the word *community* and its concept. It was an enchanting, heartening discovery.

Some early encounters opened the door to how different life was in a small town.

Chief

A couple of weeks after coming to Spring Green, I ventured into the grocery store. Unable to find a particular item, I found a middle-aged fellow in one of the aisles, on his knees, stocking a lower shelf. When I inquired about the item, he looked up at me for a moment before looking back toward his work.

"You're the new fire guy, aren't you?" In twenty-eight years of life, I don't think I'd ever been known by someone I hadn't previously met. I'd never been in the newspaper, and any reputation, however limited or of whatever character, had never preceded me. And now, I'd had my picture in the paper, with an article about me, and this fellow on the floor in the grocery store, whom I'd never met, knew me.

A little taken aback, suddenly aglow in all this notoriety, I stammered, "Well, yeah."

"Hmm. You're from Chicago, and you work for the DNR."

"That's right!" He'd actually studied up on me!

"That's two strikes." His tone was flat, and his look was without expression as he focused on the shelves.

Okay, then. If your reputation is going to be out there, I guess it's good to know what it is. *This is a tougher town than I thought it was gonna be*, I thought. I really didn't know how to respond, and it showed. The article in the local paper gave a minimal biography about the new ranger with the DNR, a somewhat unpopular state agency, thought to be a little heavy-handed in their regulatory responsibilities.

After a moment, he laughed a little, apparently somewhat satisfied with how he had befuddled me. "I'm Jerry. It's in the next aisle, about where we are," he offered to my previous question. I stumbled off, thinking about who else might be in the next aisle and if I needed to prepare to defend my reputation.

I didn't know it then, but the fellow on the floor was the owner of the grocery store and would become a friend and someone in town I would grow to greatly respect. Despite the apparently gruff exterior, he was a kind man and approached his presence in the community seriously. He headed the volunteer ambulance crew, an arm of the fire department with whom I would work in my professional role. I later became a volunteer EMT, working under his leadership; I always called him Chief. He was cool and clear-headed under circumstances of extreme stress and acute need and directed people and resources with a calm, compelling style. As a superior leader, it is hard to calculate how many lives were saved and how many folks experienced enhanced medical resolution because of his decisive actions.

His presence in Spring Green would make my experience there encouraging and would help me appreciate the nature of community in a small rural town, one of the most expansive and continuous epiphanies I would undergo there.

We never became particularly close, but we saw things somewhat similarly, and we both had helping personalities. There was a reassurance in his friendship and commitment that I found comforting. When a kind but emotionally troubled fellow, a mutual friend who had worked for him at the grocery store, died accidentally, we shared the same conclusion that he may have finally found his peace.

The store, under Jerry's influence, was a good friend to the food pantry I would lead years later, providing free and discounted food and allowing us to recycle the vast amount of cardboard we generated through the store's recycling arrangement.

He was a crucial prism for my experience in Spring Green, and I have always been thankful for him and his friendship.

WHAT EXACTLY IS SO DIFFERENT?

So, what are the elements that make country life so different? The most obvious ones came to me first.

People in the country have more space. The lots in Spring Green are bigger than anywhere I had lived before, and a lot of people lived outside of the village where space was even more plentiful. Folks are just a lot more spread out. Wouldn't it seem logical that all that space might inhibit interaction and encourage isolation? Nope. Despite this, or perhaps because of it, the sense of "commonness" or unity is far greater than anywhere I'd lived.

This struck me, especially when I encountered the existence of the volunteer fire department. Even just the words didn't fit together for me. *Volunteer. Fire department.* In Chicago, such an idea was inconceivable. Chicago had unions and fire personnel bargained as a group for their compensation. They were willing to get tough, even to the point of threatening job action (or inaction). The bargaining that took place was with the city government, which employed them and which served as insulation between those in the fire service and the people. But in Spring Green, those in the fire service and those in local government and the people being protected were often the same folks; they lived across the street or down the road.

And as I was to learn, these volunteer fire people spent an enormous amount of time and personal energy training for fire protection work and then responding to fires, vehicle accidents, spills, and other emergencies. Additionally, a branch of the fire department, headed by my new friend Jerry, provided emergency medical services, requiring another distinct and extensive set of skills and commitment. Injuries from vehicle accidents, medical emergencies, and transports from medical care facilities, like clinics and nursing homes, fell to the EMTs.

Women on the fire side of the operation were rare, but more than half of the EMTs were gals.

The community's desire to establish a tangible connection to me became clear soon after I arrived in Spring Green. When people found out my last name was Anderson, they would probe into my relationships with other local Andersons.

"Are you related to Al Anderson, the electrician that fixes TVs?" When I replied I was not, they immediately moved on.

"What about John Anderson up in Honey Creek?" When I again responded in the negative, another Anderson entered the interrogation. Realizing that this ongoing process could consume a significant part of the day and knowing it would yield no fruit, I interrupted with the news that I wasn't from anywhere nearby and didn't have any family locally, Anderson or otherwise. That was met with a deflated mixture of mild disappointment and a hint of wariness. That didn't feel too good. But looking back, I think it was just an effort to identify the central question of a community: How do we fit you in here? The failure to connect me to an existing clan was not fatal, but neither did it provide progress in resolving the "fit" question. It just meant I had to establish and demonstrate my place from a slightly more remote starting point.

This expression of commonness, of unity, was new to me. They not only cared for each other, they sought to identify individualized criteria to rationalize why such care was appropriate, justifiable, with each individual. Those criteria generated a deeply ingrained sense of responsibility for one another. The web of life worked differently here. I saw the same people on the street, at the grocery store, at church, at the gas station, and at school. Folks in a small town live almost all aspects of their lives together. For the most part, they know most of one another's news. They share in each other's joys and sorrows and in their shortages and abundances. Our neighbor out in the country once knocked on our door and told us that their sweet corn was ready to be picked and that they had had something come up that prevented them from harvesting it all. We were welcome to pick as much as we'd like. We responded, "Thanks! We'd enjoy a half a dozen ears."

"No, no," they said. "We mean for canning. You should take a hundred or a hundred fifty ears." Huh? I don't remember how much we ended up taking, but the idea was, "We have enough, and you don't have any. So take some of ours." That offer repeated itself in future years, even when they had the opportunity to harvest it all themselves.

One of the things I had to learn to account for was the fact that almost any errand took a little longer than it did when I lived in the city. That was because there was invariably a measure of "nonessential" interaction associated with every task. That wasn't necessarily a required part of an exchange, but it was a natural outflow of the nature of community and the associated relationships. And over time, it became the norm for me, as it was for everyone else. After a few years, if I went to the grocery store and failed to see and speak with someone I knew, it was almost a letdown.

Another quaint feature of life in a small community is the noon "whistle," which isn't a whistle, but it does emanate from the sirens strategically placed in the village. They are set off daily in a particular cadence, up and down the "scale" once, distinct from the pattern for a fire or a weather warning. I'm sure there was a time, long ago, when it served to calibrate people's workdays, perhaps remind them to eat lunch, or "dinner," as most here call the noon meal. Today's technology has eliminated the need for such reminders, but the siren continues. It provides a sameness, a regular resetting that transcends all the other changes in how our lives are lived, perhaps salving some inherent desire for an element of constancy, routinely grounding us from our frantic existence. As the siren's sound winds up, its reassuring presence momentarily dissipates our many preoccupations. And it does so for all of us at the same time, reminding us again that we are a single operational community—together. I imagine the day is coming when the siren's purpose at noon will be deemed unnecessary, and it will cease. I'm not sure I want to be here when that happens.

Won't You Be My neighbor?

One day that first spring in Spring Green, I was out in front of our little wood-heated log house, and a tractor rumbled down the road. When the driver saw me, he quickly slowed.

"You lookin' for wood?" he shouted loudly, trying to be heard over the tractor. No, "Hello," no, "You're new here, aren't you?" *Of course I'm new here. Everyone knows that. What's the point in asking it?*

Again, this direct approach, without any amenities, took me back a little. But then, like everyone else, he knew more about me than I did about him. "Well, yeah," was my minimal but contextually adequate response.

"Well, I got some trees at the edge of a field I need to get down. You can have 'em if you want."

Never one to look away from a gift horse, I jumped on it. "That would be great!"

"I've got a wagon and a tractor you can use to haul 'em." Who was this generous guy? "And I've got a splitter."

"Great! Thank you very much." What's a splitter? Since the guy was willing to do everything but stack it, what difference did it make? I'd figure it out.

There was some muffled amusement the next morning when I needed a few pointers on the operation of the tractor. But my neighbor, Gene, didn't seem concerned. The effort garnered quite a bit of wood. Even more, it was encouraging to experience assistance from previously unknown neighbors. We had now become tangibly connected to our rural neighborhood, folded into the doings of our new community.

In the years ahead, we got to know this dear man and his wife, Delores, too. They took to our young son, Ben, who was only a few months old when we moved in. As with any young boy, Ben was fascinated with the equipment that periodically rolled by the house. When he reached three or four years of age, Gene would ask if he would like to ride on the tractor, which forever endeared him to Ben. Soon, Ben wanted to play "Gene and Delores." "I'll be Gene, and you be Delores," he'd say to his mother. Jan was honored to be cast in such a venerated role in Ben's world.

PART OF THE FABRIC

In later years, I became involved in the ambulance service as an EMT. And after that, I ran for and was elected to the school board. Those roles felt like my "place," a need I could fill, the right and natural thing to do as a community member. More links were forged in multiple directions (e.g., on the school board with the superintendent, other board members, teachers, parents, and students, all of whom, to one extent or another, were neighbors), and I came to appreciate how the fabric of the community was enhanced and strengthened, one person at a time.

The nature of my forest fire work was public service, so I got to interact with twenty-plus other volunteer fire departments in other communities. That sense of common purpose grew in me and spurred a desire to help with all sorts of incidents. When I was close, I would help get information in advance of the fire department's arrival at a structural fire, saving precious seconds. I would do the same at a vehicle accident and, if needed, would help with traffic control. It was a part of this whole new idea that we all do everything together; we're community.

When people are aware of one another's needs, they find innovative ways of taking advantage of that in the most constructive ways. When fuel prices spiked in the early eighties, one savvy group directing a fundraising effort for a local need came up with the perfect prize for the associated raffle: a publicly displayed farm wagonload of cut and split firewood, as people were increasingly turning to wood heat for their homes. Raffle sales far surpassed all previous similar efforts with other comparably valued offerings.

I think this "mutualness" is how we are wired. In community, natural connections happen between people for all sorts of reasons. When we care for each other and feel a sense of responsibility for each other, it brings substance to our relationships and provides a practical reason to empathize with one another. There is a measure of satisfaction in knowing that almost everyone that you intersect with during the course of a day in such a community has, to at least some extent, a vested interest in your well-being and that they are comfortable with that.

I Love a Parade

Such a desire for connectedness leads to such rural communities doing things together that don't happen elsewhere. It was a rare community that didn't have an annual parade for one reason or another. One town celebrated Wild West Days; another had a Butterfest. Everyone who wanted could participate: schools, service clubs, businesses. Spring Green had a strong musical orientation, and one of their most popular annual parade participants was a band, not from the school (though they had one and were in the parade, too), but comprised of marching band graduates, some from many decades earlier. And they were good, too.

The biggest parade was the one at Witwen celebrating the Fourth of July. Witwen is an unincorporated village in south-central Sauk County, a straight quarter-mile or so of close-set houses along Highway E, home to maybe sixty or seventy folks. At the north end of town, between two robust branches of Honey Creek, is a long "campground," arrayed from east to west, with several scattered buildings, including the "tabernacle," nearest the road. On the Fourth of July, the big parade trooped its way through Witwen, south to north, in late morning. Units from at least four local volunteer fire departments were there every year, sometimes more. Smokey Bear made an appearance there every year toward the end of the forty-five-minute procession. But the most prominent and cheered participants were veterans, ranging from World War II survivors, older gentlemen, often in the back of open cars, to Vietnam vets, younger, sometimes scruffier, from a very different experience, usually walking. I never got to see the parade because I was always in it, but I was told the contrast was striking in a way that reflected all the change the country had been through in the sixties and seventies.

The parade route was lined several deep with several thousand people from near Highway O south of Witwen all the way through town to the campground. Most folks would stay for the activities at the campground. The only parking was on the shoulders of Highway E, Highway O to the south, and Witwen Road and Elm Road to the north. Cars were packed solidly for more than a mile in all directions.

The feature of the day was the chicken dinner, and I can remember when it was five bucks. I think they usually planned for a thousand, and the preparation to slow-cook the chicken on long, open charcoal fires began early in the morning. Sometimes they ran out. The dinner included dessert, mostly homemade pies and cakes baked by local gals, sliced and set out on Styrofoam plates laid out on tables set up earlier in the day. Depending on the weather, those tables would either be inside one of the buildings or outside under the partial shade of a big elm tree. The tables weren't monitored except to be refilled, and you took whatever caught your eye. There was a smaller building where you could get a soda, an ice cream bar, rolls of caps for cap guns, various candies, sundry toys for kids of all ages, and water balloons. There was much socializing. It was, to some extent, a fundraiser for a local church, but it had such a long tradition and was such a wholesome event for folks that the benefit feature was almost forgotten. One time, at a planning meeting for the year's event, someone suggested that significant additional financial benefit could be gained by selling beer as well as soda. That idea was summarily quashed.

There were horseshoe pits, always in use, accompanied by much partisan cheering and encouragement. In the afternoon, a series of throwback two-person games took place in the field between the tabernacle and the road, including three-legged races, water balloon toss, wheelbarrow race (one person holds another's feet up while they hand walk to the finish line), and the ever-popular egg toss, using raw eggs and covering an ever farther throw until only one team remained. A friend of mine and I were runners-up one year and would have won if I had taken off my wedding ring. We (I) broke at about forty-five feet. The field was lined with people cheering for family or friends in the races. Participation was so strong that there were often heats featuring ten or twelve pairs. And all races and heats were "called" by an announcer with a mike and makeshift sound system. It was the same person every year, and I was always astonished at his capacity to know the names of everyone in all the races. It was all very folksy.

Crowds thinned out as the day wore on, and the sounds of fun and celebration transitioned toward the sound of cleanup, involving many of the same people who had arrived early in the morning to start preparing the meal. Most folks moved on with their evening, but some were only taking a break. They returned in the evening for the "meeting" at the tabernacle, closed and quiet all day. Worship was often led by local personalities, like Elmer Childress, the beloved weather director at a Madison television station, and his wife, both gospel singers. Then a sermon of a gospel orientation would follow, often by a guest speaker from out of the area invited specifically for the event. Invariably, the time concluded with an alter call.

The Witwen parade gained such notoriety that one year,

Charles Kuralt brought his Peabody and Emmy award-winning program *On the Road* to track and share the event. I don't think they stayed for the evening's events.

What's the Buzz?

It was still pretty early in the fire day when the phone rang.

"Don't you think it's a little windy for them to be burning at the dump today?"

I didn't know who was calling or which dump they were referring to. In theory, the municipalities were supposed to clear with me any burning they did before they started. "Who is this?"

"Just someone who lives in town and is concerned about all the smoke."

"Where are you?"

"Right here in Spring Green."

So, apparently, there was burning going on at the dump west of Spring Green. I was a little surprised because Spring Green was usually pretty good about getting in touch with me before they did any burning. The west wind was supposed to continue at a fairly brisk rate throughout the day, so I decided I should go in and check it out.

As I passed the fire station, only half a mile or so east of the dump, I noticed that the doors were up and a couple of the trucks were out. But when I reached the dump, the gate was closed. The wind was strong enough that all I could see was smoke, so I parked on the side of the road and clambered over the locked gate to see what was going on. The wind let up a little, so I thought I could sneak around the east side of the fire, "under" the smoke, to see what was on the far side. But as soon as I committed to that, the wind picked up, and I found myself progressing to the north through very heavy white smoke from the fire. There was nothing to do but continue, so I persevered, feeling my way through the smoke, knowing that I would soon be in the clear. Though it seemed to take an eternity, I finally cleared the smoke, albeit coughing and with irritated eyes.

As the view became clear, I was surprised to see not only Spring Green's fire trucks but also several police squad cars, all with associated personnel. One of the squads was Spring

Green police, and the rest were from the Sauk County Sheriff's Department. They were safely to the north and west of the pile, well out of the smoke's range. The presence of all these vehicles perplexed me, and I had some difficulty trying to make sense of what was going on.

As soon as I emerged from the smoke, heads began to turn in my direction. I recognized quite a few of them, including some of the cops. Smiles quickly showed, and some of them began to laugh. After a quick look around confirmed that the fire was confined to the pile, I found myself confused about what was apparently so funny. One of my friends, a Sauk County sheriff's deputy, walked up to me with a big grin.

"Are you okay?"

Still befuddled, I responded, "Yes. I think so. I just took some smoke."

His grin broadened. "Yeah, I see that."

"What's going on?"

Another friend from the Spring Green Fire Department had joined us by then, and my cop friend said, "I'll let him explain it to you," and walked away, chuckling.

Through periodic guffaws, I was told that the county had initiated a coordinated effort in the southern part of the county to identify and eradicate wild and not-so-wild patches of marijuana and gather and burn them.

That's what the fire was.

That's what the smoke was that I had walked through.

And that, apparently, is at least part of why I couldn't figure out what was going on. As I began to wrap my head around what I was experiencing, I started to catch up on everyone else's laughter. It got to be kind of funny to me, too. Soon, it also occurred to me that I had an official emergency vehicle parked out by the gate, and I suddenly was in no condition to operate it.

Eventually, after they'd all had their fun with me, one of the firefighters drove me back to the ranger station in my truck. And soon after that, somebody from work drove me home, at which point I promptly took a long and pleasant nap. As I recall, I still got paid for the whole day. My work was a very "high" priority to me!

TRUST

Trust plays a substantial role in the life of small, interwoven communities. I discovered another aspect of this when I had to order a part for a piece of equipment I owned. The man at the store couldn't locate the catalog right then and was, therefore, unable to quote a price for the part. But to my surprise, I found myself perfectly comfortable with that and placed the order anyway, confident that a fellow community member would not charge an unreasonable price. And he didn't.

Another time I ordered something under the same price-unknown circumstances. When I went back a few days later, the store owner informed me, without my prompting, that the price he'd found was outrageously high, and so, on my behalf, he had canceled the order. He was looking out for my interests; that's what we do for each other. That's what they would naturally expect me to do for them if the tables were turned. We will be in one another's future, so it matters how we treat each another today. It's the code of the community.

Simplicity is another element of a small-town community. For most needs, there is only one place in town, if any, to fill them. As a result, there is no need to do a lot of comparison shopping. In turn, this brings an element of common sense into the town's affairs. People know their roles in the community, and they treat them with respect. This breeds a sense of integrity, born not of regulation or competition but of relationship and reputation. One fellow came to town and started an auto repair business. He did good work, but he charged city prices, thinking he'd make a bundle because of the lower costs of working in the country. He was out of business in five years because people saw that he took unnecessary and unfair advantage of them.

Regulation tends to be a little more sparse in the country than in the city. As a result, the compliance standards for how things are done are more closely related to relationships and the associated expectations than they are to regulations. It is a good thing when people do the right thing, not because they are compelled to do it but because they recognize it is the best way to proceed. That reinforces an inherent drive to be both righteous and wise. The process of identifying what is righteous and wise helps build character, and that is helpful to *any* community.

This ambient sense of fairness and honesty exerts its influence in unspecific ways. My older son left Spring Green in his twenties without having had extensive experience in conducting commerce there. But years later, when I rode with him as he conducted his equine veterinary practice, I saw the fruit of his home community exposure. Upon completing his diagnosis of an out-of-sorts animal, he would characterize an array of treatment alternatives to the owner. The first was "the least you could do," which was often doing nothing, not the best for his business, but certainly a reassuring option for the owner. The second was "the most you could do," the most aggressive approach to addressing the problem. Lastly, he offered "what I would do." By the time he got to the last option, the client was in a comfortable place, able to think clearly about the need, and not feeling like he was being "played for profit." I saw all three options selected, but most of the time, they did what Ben said *he* would do, and that was usually closer in aggressiveness and cost to "the least." People are very comfortable working with him because they trust him. The "mutualness" of small-town community is part of his make-up. And it did not come to him by being formally taught, but by example, by experiential osmosis, growing into the "place" the community had begun to provide to him, and all the implications that had for how to treat others. Its roots are in respect and grace.

Another time, I was traveling for a month in January and learned from someone checking the house that the downstairs door had inadvertently been left open. The water pump in the basement had frozen. From New York, I called my friend Jeff, the plumber, gave him the code to get in through the garage, and asked him to fix the problem. We didn't speak again until I got home three weeks later. Jeff found the pump housing to be cracked and decided it was best to replace the pump. So he did. I was not concerned that Jeff would do more than was necessary, and I trusted him in my house. He apparently had no concern about sinking his cash flow into a new pump for me, confident that I would take care of it promptly when I returned. The wheels of a community turn with less friction when there is trust and belief in mutual care.

Most people in such communities have fairly clear roles, often more than one. You work at the store, or you work at the library, and you volunteer here or there, or you work in Madison but are part of the Lions Club and help with this or that. Or you grow older but help manage activities for other seniors with bigger problems. Everyone has a place. Some have various limitations or disabilities, and the community subtly watches over them and finds niches that those folks can serve in one way or another. There are certain things that you don't help with because that is so-and-so's responsibility, and it's all they can do, so they do it, and they are the only ones that do it because that's what gives them identity and purpose and value in the community. Community roles shift with age, but that happens gradually, the opposite of the retirement party or the proverbial pink slip. Families are good and strong supporters of people as they wend their way through life. But communities help fill needed roles, too. It is rare that communities don't have their own retirement/nursing facility. Many, even most, of the residents of those facilities have never lived anywhere else. All they know is there, and communities find a place to keep them close and cared for and visited when they can't be home anymore.

In this day of electronic efficiency, how is it that this flesh-and-blood model persists? Some of it is tradition. On the fire department, for instance, a lot of sons follow their fathers into and through the service. And there are family businesses that are handed down from one generation to the next. But traditions die unless they provide a tangible payoff. This community structure provides, generally, a place for everyone, a framework, reasonably well-defined relationships among people, and a sense of belonging. That generates a deep-seated sense of security. And so, it persists.

No Problem!

Another symptomatic community experience for me involved a time, early on in Spring Green, driving home when I had to pull over to the side of the road for some reason. I had not been there more than a minute when some unknown soul driving past slowed and stopped to see if I was all right and if I needed assistance. When I responded affirmatively, they moved on. Once I got past a brief sense of being intruded upon, I was astonished. And as I was to discover, this was not an isolated occurrence.

As I settled into this new way of life, there was something very empowering about it, and I quickly grew comfortable with that way of thinking. Who wouldn't? It became natural for me to stop by any car pulled over on the side of any of the town roads I took to go into Spring Green to ensure all was well. Sometimes, this was just an exchange of thumbs-up signals. The treat took place when someone, especially someone from out of town, even better, from the city, was having a problem, and I got to show them how we do things out here in the country.

On one crisp fall midday, I came upon someone out on a town road, a couple of miles from Spring Green, with the hood up. Wearing a nice pair of slacks and a light-colored button-down long-sleeved shirt, standing with his hands on his hips, he didn't look like a match for any problem under the hood. I slowed and stopped, lowered the passenger window, and asked if he needed help.

After a rather dismissive explanation, which I essentially ignored, I said, "Jump in. We'll go find Joe. He knows how to fix things." A bit taken aback but faced with limited alternatives, he got into the passenger seat. After introductions, a little uncomfortable for him, I asked where he was from.

"Milwaukee," he offered, in a tone reflecting a comfort that identifying himself as a resident of a city of a half a million people wasn't revealing a dangerous level of personal information.

"Oh!" I affirmed, recognizing the same awkwardness I'd experienced when I first came to Spring Green. "I grew up in Chicago."

That revelation brought palpable relief, soon followed by a quizzical look. *Chicago! Then you know better than to pick up strangers*, was the obvious thought. He was a little more relaxed as we headed into town.

When Joe's garage was obviously closed, he sensed failure. "No worries," I said cheerfully. "It's a quarter to twelve. Joe's probably across the street eating dinner." I jumped out, leaving the engine running (with a stranger in the car!), and went to consult with Joe. Returning shortly, I told him Joe would be out soon and would meet us, with tools, at the car. The pace and ease with which all these arrangements were made was head-scratching to this new guest with car problems, and there was a sense that he was trying to catch up with the quick pace of all the interactions.

Within ten minutes, Joe arrived, a simple repair was made at the side of the road, and he released the visitor with a, "There you go. All set." It was a statement of closure, an invitation to continue on the previous, temporarily interrupted path.

Still a little incredulous, he countered with, "Well, what do I owe you?"

Now Joe, too, took some pleasure in this ability to confound with friendly assistance, so he had almost forgotten that this was his living. "Oh," as the idea of compensation interrupted his thought process. So as to satisfy the interloper's need to pay something, "How 'bout ten bucks?" Now, city folks are used to hourly rates, big ones, with a one-hour minimum charge. But ten bucks later, he gladly shook a greasy hand, and we had a new friend with a "you're-not-gonna-believe-this" story to tell when he got home. And our day was made. Joe and I forged a link in our relationship that fits into being a community to others who may not have previously experienced it.

THE FARM

One cannot adequately address the idea of Midwestern rural life without speaking of the family farm. Farming is a community within a community, one that shares much with other components of rural society but has some unique additional elements of its own.

Surely, a farm is a bit like a mistress, a demanding one, but one who is faithful. When it works best, she can function in that role for an entire family, requiring much but yielding life, and richly so.

I have never farmed, and I have only gardened with marginal success. Surely, there is a heady sense of success when a plot yields something good, clean, rich in color and texture to eat that radiates a sense of having produced something valuable with one's own hand. Of course, there is also the small matter of the miracle of growth, about which science has enabled us to understand how but never why. Most people who grow things understand and appreciate that, at some level, and recognize that farms are about working with the earth to grow things.

Most farms are generally similar in appearance. They tend to be compound-like for operational reasons—a house, a barn, and any number of assorted outbuildings, all different from one another, constructed one at a time for a particular need of that day. Approaching on a driveway, one has the sense of entering into a place of purpose. Security tends to be light. Equipment and tools are out in the open or in buildings with open doors or no doors at all, and many a tractor key, if there is one, has never left the ignition. Ever. But there is a sense of sanctity there, that everything is as it is supposed to be. In spite of the dirt, the well-worn equipment, and the various aromas that always fill the air, one senses they are in a hallowed place, one with a vital purpose and a time-honored process to accomplish that purpose that shouldn't be unsettled. If I were looking for someone, especially in the middle of the day when any work being done would likely be in a field, I would search through the barn, silent except for the occasional cooing and flapping of pigeons. I felt a need to tread lightly, almost tiptoe; this was a place where crucial activities take place.

Ongoing milking activities are usually indicated by the sounds of a country radio station emanating from the barn, spilling out either country music or the folksy programming associated with a small local community. Weather and commodity prices are an essential component, along with news of local activities, such as school sporting events, community activities, and who died. But as insignificant as it may sound, a good programmer at a country station has a pulse on the information that a community wants to stay connected with, and those are its components.

The farm, especially a dairy farm, provides a context for the most essential components of life, a pattern of the day and of the season that harks back to Leviticus. The shape of the day teaches about the "regularness" of life, the value of discipline and faithfulness to the routine a milking herd demands. Those patterns are essential because, without them, the entire operation of the farm begins to decay.

The seasons, too, make their urgency felt. The timing of our school year (and the three best reasons to be a teacher: June, July, and August) is tied to the need for fieldwork to be accomplished in the growing season. Again, activities on the farm are dictated by the sun: planting, cultivating, and harvesting.

A share of what grows is sacrificed to the raccoon, the cutworm, and the turkey. That is a price for plying the wild earth; it is understood, accepted, and constrained. At a glance, that can seem like a violation of justice. After all, it was the farmer who does all the work. But there is a lesson of coexistence, perhaps a small price for the unknown "why" of all the earth provides. The biblical concept of gleanings points in that direction, to not harvest too efficiently so that the have-nots willing to work can salvage the residue the harvester misses (or leaves).

A young person of workable age (which is earlier on the farm than it is in the rest of our world) learns the need to say no to many things because of chores. When a student is liberated for a school effort, it is a family effort to make it so. Not only is there transportation needed from the remoteness of the farm, but someone has to cover the foregone duties. That provision can occur through simple grace or a trade of duties with another family member to be performed at more amenable times by the one with the opportunity. Wrestling requires a combination of strength, commitment, perseverance, and a willingness to fight through difficulties. There's a reason so many great wrestlers come from farm backgrounds.

On the way home from somewhere, I had stopped at a farm to speak with the owner about a tree planting he'd inquired about. It was close to eight o'clock at night. I pulled into the yard as a tractor pulling a full hay wagon came around a knob from a back field. I waited as he pulled up and stopped, stepping out of the door of the cab on the other side. Around the front of the tractor came a skinny little fellow who couldn't have been more than ten years old.

He purposefully walked right up to me and very politely but very authoritatively asked me, "Can I help you?" In that one question and the way he asked it, so much about what a farm imparts became clear. This was his farm, too. His politeness in responding to my presence betrayed his appreciation of the blessing it was and the need to treat others well because of it. But it also communicated a sense of hard-fought ownership rooted in his sacrificial response to its demands. He understood the justice of the farm, what it required, and what it provided. That simultaneously led to a servitude to the land's demands, a dependence, but also an appreciation of what the land could and would provide in exchange for faithful service. Some people never learn that principle of justice; this kid was ten and already got it.

A farm's dividends take place on levels our culture has largely forgotten how to appreciate. There is the quid pro quo of a habit of a hard day's work for a field full of corn, a tank of milk, a barn full of hay. Most everyone appreciates that exchange. But farms grow discipline, too. Cows need to be milked every morning, whether you feel like it or not, whether you're sick or not, whether something hurts or not, even if it's Christmas. And when it can't be done, because of an accident or a serious illness, neighbors jump in and help. They know what's at stake, and they understand what has been invested. One of the loveliest gifts I witnessed was a family offering to cover the Saturday evening milking for another family so they could fully celebrate their daughter's wedding without the interruption of chores.

Even more, for a young person, a farm is a base point, a dependable fallback that provides essential truth to anchor their life. There is a structure here. Your work here is valuable. Your presence is essential. Your role is clear. Your future is secure. Those messages and their dailyness are a powerful counterpoint to the confused message of the world. For the John-Boy Walton types, the world beckons, and rooted with the confidence of their time on the farm, they venture out and change the world. But for others, the charismatic draw of the farm, the allure of its ongoing exchange is a more than natural draw. They remain, marry a substantial and like-minded partner, and the cycle continues. The farm is more than a vocation. It presents a comprehensive way of living and coexisting. Its reassurances are constant and dependable. The rising of the sun, the changing of the seasons, and the cycles in animals' lives that lead to the sacred deer hunting season present a reassuring encouragement in tough times and a grounding when they're good.

The gradual changing of hands of a farm from one generation to the next has lessons of its own. Mostly, it goes well, with the gradualness that time offers providing for the growth of discernment and the smoothing of the rough edges that surface at such a time of loss for the previous generation. "Farm women," a term sometimes used derisively, are a rich source of insight and wisdom in a farm operation (to say nothing of their labor). This lubricating balm makes farms and transitions operate more gracefully.

The family farm is becoming a thing of the past, sacrificed to the efficiency of the corporate agricultural operation. Economically, it may be a smart move. But there is so much more to a family farm than economics. We have not yet fully grasped what we are forgoing in allowing the family farm to waft into memory. However, a careful look at our culture reflects that loss. The focus on economic mindset shows itself in other unhelpful ways, too. We've lost our daily rhythm with the sun, more damaging than we know. And we're losing the appreciation of the essence of the value of a day's work, which the farm so inexorably demonstrates. And we are the worse for it.

Farms play a large role in the dynamics of rural communities. Some aspects have a more romantic element, and others are more pragmatic, especially for the family. Every young person reaches a point when they begin to decide what they want to do with their lives, what sort of work they want to do. For most kids, and a higher percentage in more urban settings, the work examples that their parents demonstrate can be difficult to assess from a practical standpoint. The professional parent has minimal opportunity to share with a young person what that work might be like on a day-to-day basis, even if the youngster is generally attracted to the sort of work the parent does.

Working a farm, on the other hand, tends to be a family affair. Why is that? For one thing, the office is local. The home and the work buildings and the acreage, where the work is done, are all in the same place, "co-located" in city jargon. The workload is not evenly distributed throughout the year. Planting season in the spring and harvest time in the fall can require almost around-the-clock effort. That compels the participation of family members in ways, and at ages, that simply don't take place in most other sorts of businesses. As a result, farm kids get to enjoy a kind of trial run at the family business in a pragmatically experiential way so they can make a better-informed decision about whether they want to carry it on. Some do and some don't, and that's okay. Regardless, they get to witness a good work ethic, with a good opportunity to develop their own. They also gain insight into the food cycle that provides a strong dose of reality about how the world works, that food doesn't come from the grocery store. The opportunity to work in the dirt, on the ground, with the land, with one common purpose provides intangible and invaluable experience that serves life in ways that aren't fully appreciated. There are good lessons in mutuality of support among the farm community, a lesson that is more foreign to other competitive business activities.

Grounded

The fire had been mostly inconsequential, quickly suppressed by the fire department. Since it was on one end of my fire area when I had been on the other, I wasn't able to respond at the time of the fire. It sounded like a pretty straightforward fire from a trash pile, but I needed to have a look.

It was a heavy gray day, dry, but the sky seemed pregnant with moisture. An indistinct driveway seamlessly morphed into the woods on either side. When it ceased to be a driveway, there was a small, squarish one-story house. Yardless, it was not in overt disrepair, but it had a shabby feel. The river wasn't visible from the house, even in spring before the leaves came out, but I could see the light through the tops of the trees in the distance where the swamp gave way to open water.

Quickly spotting the burn pile and the minimal area where the fire walked into the woods, I found the evidence to be consistent with the fire department's assessment. The homeowners were characterized as kind, so my last step was a quick conversation with them to discuss regulations and consequences. I knew nothing about who lived there, but that was about to change.

A knock on the door beckoned a middle-aged woman. When she opened the door, my olfactory was assaulted by the intrusively sour smell of cat "activity." I began to take in the scene in front of me. The woman was humbly dressed and looked to have some hard years behind her. Among the too numerous cats were a generous handful of children, all barefoot, tired clothes, some hand-me-downs undoubtedly on the second or third child, stained, if not dirty. Most appeared too young to be in school, but a couple didn't, and this was a school day. The motion of cats and kittens seized my view in almost all directions in the house; there were surely a dozen or more. The next thing that struck me was that the floor was dirt. Not dirty. Not needing to be vacuumed. It was comprised of dirt. Impressions can come quickly and absolutely sometimes, and this was one of those times. My thinking had gone to a new place. The issue related to the fire suddenly seemed like barely a secondary matter. At the same time, I really didn't have any idea how to respond to what I saw. I was looking at a caricature of how the fabled hill people of Appalachia live, one I didn't think even existed. My senses were overwhelmed in the same way they had been on a couple of ambulance calls to grisly automobile accidents. So much was wrong, so much needed to be done, and I didn't even know how to start thinking about the situation, let alone do anything effective to help.

As my purpose came back into my mental focus, I knew there would be no point in passing on the suppression bill from the fire department. I had a brief conversation with the woman, patted a couple of the kids who curiously came over to see me, and left.

Walking back to the truck, I had the sense of having stepped back in time. Disoriented, I found myself looking for small, reassuring things to confirm that I was still in the late twentieth century. My back-in-service call to Louie was more reassuring than it should have been, confirming my enduring connection to the real world.

It can be so easy to assume that most people's life experiences are within a degree or two of similarity to our own. Experiencing, even briefly, a home that could have been from a half-century previous was a potent reminder that not everyone's life is like mine. That experience moderated my perspective on comfort, possessions, and how I contemplate the notion of security, both immediate and longer term. It also reframed the importance of my work. That tempering and the images of that family have maintained a stark position uncomfortably close to the front of my memory ever since.

Barred From Disclosure

It all started just for fun, without any notion of the awkward situation it would generate.

My young son, Ben, and his friend, Mitch, had been playing at our house all afternoon. His dad, Kevin, came by to pick up his son, and we were all talking when I remembered something that belonged to Mitch on the far side of the house. As I retrieved it, it struck me as a fun idea to try the barred owl call I had been practicing, just for a lark. I did so from the other side of the house. Immediately, I heard Kevin, a knowledgeable outdoorsman, excitedly say, "Oh! Did you hear that?! Boys, that's a barred owl. You don't hear them in the daytime very often." He was serious. He began to explain their habits and habitats, the schoolings of an expert. Now what do I do? I was pretty sure Ben knew it was me. And we both knew that a barred owl in the daytime was even rarer than Kevin knew.

When I returned, Kevin queried, "Blair! Did you hear the owl?!"

"I heard something." It was a true statement. I forced myself not to look at Ben or my younger son Pete, or we'd all lose it.

It was best just to let the whole thing play out. Kevin would grow in his legendary reputation of having an encyclopedic knowledge of local wildlife. We would have a fun family secret. And I had the satisfaction of knowing that my imitation of the owl was good enough to fool somebody who was well seasoned in such things.

Healing

> "Old man, take a look at my life, I'm a lot like you,
> I need someone to love me the whole day through."
> —Neil Young, "Old Man"

It had been several years since the old man had lost his wife to cancer. It may or may not have been a particularly close relationship, but they had been together a long time, and it seemed they had grown to depend on one another more than they realized, perhaps more than he wanted to acknowledge. Maybe he tried to take her passing in stride, which, of course, was not possible. Maybe there was some underlying anger that led to a level of bitterness. Whatever happened, it led to isolation. And few good things are bred in isolation. He approached the world with a perceived sense of disrespect, and that worked its way into almost all of his interactions.

One Saturday morning, we did what's called a mock fire, which is a pretend fire that gives different agencies the opportunity to work together in somewhat realistic fire suppression circumstances. The setting for this one was a pine plantation generously mottled with lots of houses. I was responsible for a squad that included four trucks from two fire departments, one of our tractor plows, and another ranger. We were to "protect" a cul-de-sac where there were several houses, and the fire departments began to unload their water, not directly on the houses but into the woods, to simulate the depletion of the water they would discharge on an actual fire. That was essential because part of the exercise involved establishing a water point that would be used to refill trucks. That water point turned out to be a significant bottleneck on which we focused significant discussion after the exercise.

A few residents were out, cautiously observing the activities. But as we were doing our water exercise, the old man came out from his house, charged up to one of the firefighters, and demanded, "Who's in charge here?"

The fireman hesitantly pointed at me, and the man charged in my direction. "Are you with the DNR?" he yelled, more accusing than asking.

"Yes, I am. I am the ra—"

"You sons of bitches were supposed to get me a permit so I can put in a new well here. Your guy promised to get me the paperwork in a week. That was a month ago. I've called him five times, and he still hasn't called me back."

Whoa! *Hey, I'm just here for the fire exercise!* I thought. *This guy's been loadin' up for something more.* First, counsel to self: *this is not about you, and it's not about your heritage.* Despite what you see on TV and the news, most people are good and kind and open. There are some serial antagonists, but not nearly as many as we are led to believe. So my guess was this guy had some basis for being irate. I chose to take it as a challenge.

"Well, that's not good serv—"

"You sons of bitches take all our tax money and don't help us out at all," he interrupted me again. "You sit in your offices and do nothing while those of us out here that have needs get no help." Then, in a more victorious tone, he said, "Maybe one of these days, I'll just quit paying my taxes!" With the fire of his anger, he smugly awaited the impact of his threat. By this time, a couple of his neighbors had strolled over to listen to the exchange.

"I'm sorry that happened." I finally got to finish a sentence and established an interactive connection.

My apology was not what he expected. "Well, I just get tired of people not doing what they say they'll do." That didn't take long. The steam was already beginning to dissipate.

"I hear you. That bothers me, too. Water resources are not my area of responsibility, but I know the guy who does that. Do you want me to talk to him?"

"Well, I don't know. I suppose I could give him another call myself." With just a precious few empathetic words, his aggressive negativity had been defused.

Maybe I even had a chance to make some positive inroads. "Okay." I paused. "Did you get the note we sent out about this exercise today?"

This time, one of the neighbors stepped in and spoke up. "Yeah, I saw that, but I don't really understand. Why are you spraying water?"

As another neighbor leaned in, I responded to the question and explained the whole exercise, as well as the benefits that would accrue and how they would help in the event of a real fire. We all had quite an exchange.

"I didn't think you guys worked Saturdays."

Thank you for that segue. "When there's a learning opportunity like this, we're glad to work a weekend."

When we were done, I gave the old man my card and told him to call me if he didn't get his well problem resolved with the DNR and that I would try to help him. He looked at my card, looked up at me, and stuck out his hand. I had a new friend.

Listening seems to be a dying art. But it is one with enormous potential, especially in places where it has been long absent. Sometimes people experience frustration and just need the opportunity to have someone hear what they think. If one has the capacity to take the initial barrage, there is a great opportunity at the other end to make a good connection.

A few months later, I was back in the same area, so I took a chance and stopped at this fellow's house. "I never heard from you, so I figure you got your well problem fixed. Yes?"

He scowled at me quizzically. It took him a minute to place me, but when he did, his whole expression morphed into a smile. "Oh, yeah, I got it worked out. Thanks." After a pause, "Did you guys learn what you wanted to when you were out here with the trucks?" he reciprocated. He had taken on a whole new countenance.

"Yeah, it was a good exercise. They are a lot of work, especially for the volunteers from your local fire departments. But everybody wants good results. And we'll be better now because of that exercise. Thanks for cooperating."

"Yeah. Anytime. We're always happy to work with you DNR boys." Say what?

It's a good day when you can change a mind.

LAW ENFORCEMENT

BEING A COP

Some law enforcement stories are entertaining, but they illustrate a greater truth. I wasn't a cop, at least not in the sense of addressing law enforcement issues full time. But even what little I did required me to approach situations, when they occurred, from a negative perspective. That is, if I was going to ascertain a truth that was likely to cost someone some money or worse, status, I needed to approach the situation with the assumption that something was going to go wrong or that someone was going to try to get me to believe something that wasn't true.

Maybe engaging in a situation with suspicion doesn't seem like a big deal. But there is something in the way we are made that naturally wants to believe what we are told. If you don't think dubiousness is something we learn, that we grow into, then look at a child. Or look at one of the intellectually challenged folks in our midst. There is a lovely, wide-eyed trust there, and when we are with them, we are aware that we need to treat them delicately, as something special, even sacred. And when we interact with them, it is somehow we who are enriched.

On the other hand, there is a segment, an element of society, that tends to repackage or bend or avoid the truth to facilitate their corner-cutting lifestyle, and they can be quite comfortable expressing those variations. Many, maybe most, people have very little interaction with folks from that segment, at least that they're aware of, and are unaccustomed to routinely needing to evaluate what people say for its veracity. In my job, I learned to look for inconsistencies, to assume what had most likely happened, and to evaluate carefully and aggressively any information presented to me that was not consistent with that. Occasionally, I was proven wrong, and that was okay; truth was the objective. But as my experience grew, I was usually right. And I got pretty skilled at steering (manipulating?) people around the "bends" in their stories to get them to arrive at and express the truth. But in a given situation, after that process of extracting truth was done, my job allowed me to get in the truck and drive off to do something completely different, like check a fire, meet up with a fire chief, connect with a colleague on something, write a report, something that allowed me to "wash my hands" from the wrestling match with the deceiver. I could climb down from my suspicious mindset and interact with people I trusted, people who had no reason to try to deceive me. I could rinse myself of being guarded and get off that path that can lead to deep-set cynicism.

Sometimes we hire people to come to our homes and spend a couple of hours a week doing some of the jobs that we would prefer not to do because they're dirty or unpleasant or because we'd simply rather be doing something else. Those jobs result from physical messes that are a product of the natural course of living: dust, dirty floors, and all the fun issues associated with bathrooms. They just happen without intent, covert objectives, or malice.

There is a somewhat parallel issue in our culture, our society. Stuff happens on the streets, in our lives, sometimes in our homes, and we generally don't want to have to deal with it. So we hire police to do so. We ask them, on a *full-time* basis, to address the more unpleasant,

Some of the various badges. Clockwise from lower left: A standard badge, worn on a uniform shirt or around the neck, as displayed; a fabric badge on an in-the-pocket mounting; a centennial badge, marking one hundred years of forestry in Wisconsin, not used in an official capacity; a pocket badge, for when a displayed badge is not essential.

INTERNATIONAL ASSOCIATION OF ARSON INVESTIGATORS

CERTIFIED FIRE INVESTIGATOR

Hereby Certifies That

Blair W. Anderson

Has successfully demonstrated the ability to meet the standards for certification in the Certified Fire Investigators Program and is hereby recognized by them as a Certified Fire Investigator.

_____ _____
PRESIDENT   CHAIRMAN CERTIFICATION COMMITTEE

Certificate Number   25-052

Expires on   December 1, 2000

Because of my involvement in so many cases that went to trial, I was able to accumulate sufficient time as an expert witness that I could sit for the Certified Fire Investigators Test. Though it was about all aspects of fire, including structural fires, I studied extensively and passed it, becoming only the second DNR law enforcement officer ever to accomplish that certification. While it didn't have much practical application, it did make it easier to be recognized as an expert witness in court proceedings. And at the annual meetings of the Wisconsin Chapter of the International Association of Arson Investigators (a surprisingly well-attended educational opportunity), they printed my name tag with a "CFI" on it; it was prestigious but a little embarrassing, too. After I became chief ranger, I was less involved in field fire cases and was unable to renew the certification (or attend the meetings).

distasteful, illicit aspects of what happens in society and the deceitful people who are often involved. We call cops when somebody has done something wrong, when someone perceives that someone else has violated their rights, or when the course of events in a community has gone outside the range of acceptable or tolerable. We ask them to be in that suspicious mentality almost continuously as they do their jobs.

Think about it. I found it challenging to my perspective and my attitude to have even a small diet of those sorts of things in a much less dangerous (humanly speaking) arena. How does being immersed in such a world on a continuous, full-time basis impact a person? Unlike me, full-time cops don't get much relief from that mindset in the course of a shift. Even more, the tasks we ask them to do are so dangerous that we give them, literally, the power over life and death to be able to accomplish those tasks as needed. If one is not careful, that sort of life can lead to a contemptuous outlook, a dark perspective, a poisoning of the soul.

No one gets a free pass, nor should they. We understand that the guy who pumps the septic tank has a filthy job. And we all get that his truck and his clothes and the smell of his world are nasty. But perhaps it would be good for people to better understand and appreciate the nature of what we ask cops to do. It's nasty, too, but in a more aberrant, internally twisting sense. We have fun picturing them sitting in squad cars eating donuts. But that's a criminally incomplete, if not usually inaccurate, picture.

And like the mop-up on a fire, the dirtiest and most unglamorous part of suppression, the work police do has to be done both effectively and with integrity, or the life of our society starts to go off the rails. It's asking a lot of a human being.

A Star is Born

"When you believe in things that you don't understand, then you suffer."
—Stevie Wonder, "Superstition"

The Wisconsin River is a mecca for all sorts of soul-restoring activities. The open water provides opportunities for many forms of recreational experiences. But its mysterious backwaters, its primordial sloughs, seem to draw those who, for whatever reason, prefer to maintain a more secluded existence. They see the river bottom as a sanctuary, a robe of protection. Where the swampy lowlands are more extensive, habitations of a curious sort can exist. They may not always comply with local zoning requirements, but whether by their location or the nature of their inhabitants, no one seems too eager to compel compliance.

I received word of a run one of my fire departments in Dane County made on an otherwise insignificant fire. I got the address from the Dane County Sheriff's Office, who dispatched them, but I was unable to connect with the fire chief before I had an I'm-right-here-anyway opportunity to stop and take a look, so the investigation took place without the backstory. That always inserts a kernel of discomfort, especially so at this location, which was close to one of the swampy areas of the river north of town. It was edging toward evening.

The driveway off of the county road was at least a couple of football fields long, and a few curves enhanced the sense of distance from the highway. After calling in the address to Louie, I arrived at a tired-looking trailer with a most peculiar entryway. Perhaps forty feet long in its path to the trailer, it was comprised of sheets of various used materials, plywood, corrugated roofing, fiberglass, arranged with an upright sheet on either side, held up by fence, posts, or pipes, or whatever, with another sheet laid across the top. It had the feeling of a cave entrance, and apart from the occasional translucent sheet of fiberglass, it was dark. With each cautious step into this ominous structure, the relative safety of the open air and the skies was one step farther away.

At the door, I paused, assessing the merit of disturbing the being inside who had identified this entry as worthy of all the effort to construct it. A few seconds after tentatively knocking, I heard the knob turn. The door slowly opened, but only a foot or so. He was five feet eight inches, maybe forty-something, very slender, disheveled, wearing a heavily worn flag T-shirt, deeply faded jeans, and barefoot. Moderately long sandy-colored hair was neglected, and his longish, unkempt beard and weathered complexion completed the look. But his eyes were sharp and penetrating, almost irresistible. Charles Manson came to mind.

He didn't say anything but looked at me expectantly, a silent "What?" His appearance had distracted me, and it took me a second for my mind to route back to the purpose of my being there. I explained that I had heard about a fire there and wanted to take a look at it. He simply nodded back and to the right and closed the door. We apparently shared a common interest in limiting the extent of our interaction.

I made no hesitation in exiting the tunnel, resisting the urge to run. Relieved at being out in the open, I attentively walked around the trailer to a large field of short-cut grass, toward one far corner of which was a brush pile that appeared to have been burned more than once. I spotted a very small, burned patch in some longer grass at the edge of the field, no more than one hundred square feet. I walked around it cursorily, looking for something significant to convince me it was caused by anything but an ember from the brush pile. When nothing jumped out at me, I was ready to make my exit, concluding that a return to the trailer for another "discussion" was unnecessary.

But as I started that way, a straight line in the grass, burned black, a couple of inches wide, caught my attention. I followed it to its end, a few feet in front of me. It was an old burn, almost difficult to make out. When I turned around, I realized it continued straight in the opposite direction from where I had first spotted it. I tracked it to the opposite end, maybe twenty-five feet away. From that end, I saw that another line, similarly burned, cut back to my left. *How odd*, I thought. Engrossed, with my curiosity thoroughly aroused, I followed that line roughly the same distance to its end. Again, another line cut back off of it, again to the left. Still mystified, I followed that one, crossing my original line in doing so. Finally, I took a broader view of what was around me.

My eyes traced each line to the next until they had taken them all in, completing the figure. Though my mind wanted it to be something else, my eyes confirmed its shape. Even in the fading light, a heavy darkness came over the field. I was standing in a carefully laid out and meticulously burned pentagram.

Enough investigation. The fire was out, and I knew what caused it. This was hostile territory of a different sort. I strode purposefully to the truck, again fighting the instinct to run.

As I accelerated down the driveway, my mind went over how

much work it must have taken to lay out the design so accurately in the grass, with all its perfect angles, and then to burn each of those lines so uniformly. The mind that conceived and executed that effort was not one I wanted to engage. It was especially comforting to hear Louie's voice on the radio.

When I spoke with the chief later, he asked if I had been out there. When I told him I had, he just shook his head and forced a grim smile.

"Pretty different, huh?"

I was back to my investigative mode. "Did *he* call it in?"

He shook his head. "Anonymous. But I'm betting it was a neighbor that didn't want to be identified."

"I can see why. What's the story?" I wondered aloud.

"Nobody knows too much about him. I think he's a Vietnam vet with some bad memories. He keeps to himself, and no one seems particularly interested in engaging him."

While I felt an enhanced empathy for him, it was not enough to motivate a return engagement. The zoning folks must've felt the same way.

Out of the Mouths of Babes

"And a little child will lead them."
—Isaiah 11:6

Some fire situations include an eclectic mixture of human elements. This one occurred less than halfway through a spring season one year. It brought together a violation of the ninth commandment and a measure of redemption delivered by a child's precociousness.

The woods were beginning to dry out but weren't at any real risk of becoming the long-running, heavy mop-up fires that can come later in spring. At the north end of the county, thirty miles away, I heard a page for a grass fire east of my home station. Knowing there were no oceans of grass in the area to breed a large fire, I dismissed responding immediately, as I was occupied finishing the mop-up of another fire. Sure enough, within a half hour, the fire department that ran on it was headed back to their station.

It was late in the day before I returned south, so I resolved to investigate the fire the following day.

The next morning, I went to their fire station, let myself in, and got their run sheet, which had a variety of pertinent information on it. Some departments completed these more thoroughly than others. When they are well done, run sheets usually have some cause information that is helpful to me when I visit the fire site and speak with the people involved. Atypically, this particular report was a little sparse. But it had been Sunday; maybe they were in a hurry to get back to their homes. One piece of information the report did include intrigued me. The property owner was an officer on the fire department.

I drove to his rural property a couple of miles out from town. An acre or so grass field had burned as well as a brush pile in one corner, near the town road and the driveway that led to the house in the rear of the property. Based on the winds from the previous day, it was a pretty safe guess that the fire had started in the brush pile. I spent a couple of minutes at the end of the driveway looking at burn patterns and confirming my suspicion of the pile as the source of the fire.

Hank, the fire department officer who lived there, had apparently seen me, and as I drove the hundred yards to the house, he came out to greet me.

"Hey, Blair. I never thought you'd be comin' here to check out a fire," he greeted me, laughing. We knew each other and got along well enough.

I chuckled back. "I can imagine." I turned toward the brush pile, where we both knew it started, and asked, "What's the story?"

"Well, I don't really know." As he started to respond, I sensed something was a little askew. I couldn't put my finger on it, but he seemed more uncomfortable than upset. "I wasn't home. But there have been some bikers around. I've never seen 'em until recently, but I ran into them last week at the end of the driveway. We didn't have words, but it wasn't all that friendly, either. All I can think is that it was them that set the pile on fire. I'm not sure."

"Huh. Why would they do that?"

"I don't know, unless, like I said, we didn't hit it off too well the other day, and they were tryin' to send a message or something like that. I don't know."

The "I don't knows" felt like disclaimers for everything else he said, so I decided to press him. "Wow! You think they started that pile in your front yard in this dry weather with those winds and your house right here? You must be furious."

"Well, yeah," he stumbled, his anger awkwardly delayed. "I hate to think what could have happened," he blustered, a little too late.

"Have you contacted the police?" I asked, provocatively.

"Not yet. I'm going to figure out who those guys are first."

Something was wrong, but I still couldn't be clear on what it was. At that moment, the cutest little pixie of a girl, maybe six, burst out of the house and came skipping toward us. She stopped when she reached us and reflexively grabbed her father's hand.

"Hi," she said to me, overflowing with curious joy. "Who are you?" she asked most engagingly.

Crouching to get on her level, I said, "Hi. I'm the forest ranger. I'm here to learn about the fire you had yesterday. That must have been pretty exciting, huh?"

Her eyes grew wide. "Oh, yes," she exclaimed, aglow with excitement. "Daddy lit the brush on fire, and it got away," she giggled.

Sometimes the hardest thing to do is not grin. I glanced up at Hank, then back to the girl, and made the most non-evaluative response I could think of at the moment. "It's a good thing they got it out before it got to the house."

Poor Hank. In the flash of a moment, he had a crucial choice. Does he correct his virtuously honest daughter on what she said, which we all knew was the truth, and face the awkward, morally corrosive consequences with her, or … ? I, too, had a choice. It really boils me when people lie, and it's even worse when it is a fire person, a colleague, of sorts. The little voice of vengeance in me wanted to call him out on his lies right then and there. But the call to self-control prevailed, I'm humbled to say.

Hank and I locked eyes. Still holding his daughter's hand but looking at me, he confirmed, "Daddy made a mistake."

Good for you, Hank, I thought. Lying was wrong, and he knew better. But he chose wisely when the stakes were ratcheted up, when it mattered.

Looking back at the girl, I said, "We all make mistakes. It's nice to meet you."

She responded politely.

"I'll talk to you soon, Hank." I stuck out my hand. It was as good an ending as I could have hoped for, considering how it began.

Later, the chief called me with a half-hearted proposal that the fire department not send us a bill for this fire so their guy wouldn't be on the hook for it from me. "Nope. Can't pick and choose because of circumstances." He knew better. Billing is my call, based on circumstances. Hank got the full treatment.

Where's the Fire?

One of the subtle tensions in the job involves transitioning from winter into the beginning of the spring fire season when the first fire occurs. There was always much to do beforehand. It may be inspecting the back cans loaned out to fire departments to make sure they work or delivering fire prevention articles to newspapers or doing interviews on radio stations (this was before the days of email and effortless audio/video connection). There was also forestry work to do, and the days of early spring, before fuels dried out, were delightful times for that. But if one was not careful, one could get caught in the woods when that first fire call occurred.

Another complicating factor in my earlier days was related to money. Our local budget was charged with mileage expenses, on a per-mile basis, by vehicle, for any traveling we did. My smaller 4 x 4 fire truck was still large, heavy, and carried a relatively high per-mile rate. On the other hand, the parks folks had a very small pickup truck, a Chevy LUV, as I recall, which carried a mileage rate approximately one-third that of my truck. If they did not need it, they were willing to let me use it, and the savings were real, enabling us to pay for some other essentials with our budget dollars. The park's truck had good radio communications, at least with other DNR units, including my dispatcher, Louie. So

I could stay in touch with what was going on, even if I couldn't speak directly to other agencies.

One morning I played the transition card a little too closely. I had used the little park truck to deliver an article to a couple of newspapers in the northern part of Sauk County and was scrambling south toward Spring Green, aware that I needed to get back to my fire truck because the early rain had failed to materialize, and it looked to be a good afternoon for fire, especially in quick-drying grassy areas. Sure enough, I was just south of Baraboo when Louie called me to tell me that the Black Earth Fire Department had just been paged out to a grass fire northwest of town. As I descended from the Baraboo Bluffs on Highway 12, I could see the smoke column, and my pulse accelerated along with the truck. The red lights of a Sauk County sheriff's deputy appeared in my rearview mirror. I quickly pulled over and got out of the truck.

"Where's the fire?" came the caricatured inquiry from one of the older deputies. As it registered with him that I was the DNR ranger, I turned and pointed to the rapidly building smoke column.

"I gotta go!" I was already charging back to the truck when he stammered, "Oh, okay. Go ahead." I continued my journey to the fire. It was amusing to actually answer that question with a legitimate location.

ARREST

One of the aspects of law enforcement that always gripped my focus was the power of arrest, the power to, even temporarily, restrict a person's freedom. To empower one person with the right and authority to deny another person their freedom is a radical, albeit essential feature in a culture. It is far better granted by a society through process and criteria rather than seized through muscle and opportunity.

You're Not Going Anywhere

The need and the nuts and bolts of that struck when I responded to a fire that had very recently been suppressed. I arrived at the scene just as the fire department was packing up the last of their equipment and were about to leave. The landowner, who was responsible for starting the fire, was digging around in the area of the origin. When he saw me, he furrowed his brow, shook his head, and turned away. I thought, *This is going to be fun when the fire department leaves, and it's just the two of us.*

As their units left the scene, I told him I would need to see some identification to run a records check. With the same look of disdain, he handed me his driver's license.

"You do that. I'm going in the house." One of the things a law enforcement officer never knows is what an offender, a perpetrator, a suspect, whatever you want to call them, may have "inside the house." With the attitude he was projecting, I had reason for suspicion.

"No, you need to wait here while I do that." A smugness came over him.

Seeing I had no components of the "force continuum" typical of police (handcuffs, pepper spray, baton, sidearm) beyond a badge, he challenged me, confidently. "And how are you going to keep me from going inside the house?"

Okay, if that's how you want it. "I'm not. But if you leave here, I'm leaving, too, with your driver's license." That curdled him further, and I continued. "But when I come back in a few minutes, I will have an armed sheriff's deputy or two in squad cars with lights with me. Is that the way you'd like to handle this?" That slowed him down. He measured that idea for a moment, then leaned on his shovel, silently demonstrating his acquiescence. He was not free to leave, "under arrest," albeit temporarily.

I walked a few steps to the truck, ran the information, and everything was okay. So I wrapped up my time there quickly. He was released, and I left.

But that memory and the recognition, situationally motivated, stayed with me. It was no different than a traffic stop when a person is temporarily arrested and not free to leave. In our American culture, that notion of being under someone else's direct control is one that rankles. That authority is not to be taken lightly. I was glad I had it, and I never used it carelessly.

KXT996

I was still pretty wet behind the ears, and it was my second spring fire season in Spring Green. Heading north out of town on a sunny afternoon, I came upon a small and tentative fire that had obviously walked away from a little smoldering brush pile. I called Louie and told him I would be stopping to put out the fire. He asked if I wanted the fire department dispatched, and while I didn't think I would need them, it seemed wise for someone as new as me to request their presence. So I told him to go ahead.

The fire was out by the highway near the top of the hill north of town, in front of a humble house that was set perhaps 150 yards off of the road. Because of high-speed traffic on the highway, I pulled off of the shoulder into the ditch depression where the fire was burning to avoid traffic problems. By the time the fire department arrived, I had all the running fire out, and they poured some water on the brush pile.

A man with a peculiar gait had wandered down from the house. He was of unremarkable height or build, obviously getting on in years, but with a sharp flash in his eyes that caught my attention. Despite his awkward step, his movements were smooth and deliberate and spoke to a fitness level that suggested he could take care of himself. He had gotten pretty close to the brush pile.

As he quietly observed all the activities, I found that my truck had become mired in soft ground. That surprised me since we were up on the ridge, but as I would learn, there are a lot of springs one discovers in surprising places. I now had the opportunity to use a winch for the first time. New territory for a city boy. I wrapped the cable around a tree near the driveway in the direction my truck was facing. Using the controls from inside the cab, I quickly and effectively (even expertly!) extricated my truck from the soft ground. I was quite pleased with myself.

I released the fire department and went to have a conversation with this understated homeowner. But before I was able to engage him, Louie called on the radio to inform me that the landowner had been warned twice in the past about burning during the day without a permit. That little piece of information seemed inconsistent with this apparently mild-mannered fellow.

So, I went to speak with him and asked him how it happened. He just grinned at me, looked at the brush pile, and said, "I just had a little brush to burn, so I burned it."

"Are you aware that a burning permit is required for burning here?" I asked him, knowing the answer. He shrugged his shoulders. "Haven't you been warned about this before?"

"Maybe," he responded. Was he simply confused, or was there a trace of obstinance present?

"I'm going to need to see some identification," I told him. Without a word, the gait reappeared as we wandered quietly toward the house. We were met at the door by a kindly but slightly agitated-looking older woman, his wife, I guessed.

"What have you done now?" she bemoaned him.

"He needs to see my driver's license," he squawked back. She went into the house quietly without responding. Had she heard him? Or was she ignoring him? He followed her.

"I need to see his driver's license," I called to her, with elevated volume.

"I'm sorry about this. I know where it is. Do come in," she responded kindly, and I did. She offered a chair at the kitchen table, just inside the door. The man sat in an easy chair across the room, eyeing me as she disappeared into an adjoining room. The house was small and had entertained many years. It was worn from it but looked cared for.

The woman returned shortly with his driver's license. I had opened my small, zippered portfolio to remove a citation form onto which I could write the information from his license. As I was doing so, the woman asked what the form was, and I told her. That changed the atmosphere in the room, and the true colors of the old man appeared.

"Do you mean a ticket?" he snarled.

"Yes. I saw that you have been warned on at least two previous occasions." He got up out of his chair and grabbed a nearby crutch, moving a little more fluidly than he had before. As he approached me, the expletives began to flow, as did the volume of his voice. I coolly continued writing without looking up, something I would later learn not to do. My failure to respond to him only incited him further, and when he got close enough, he began forcefully banging his crutch on the table across from me. *This man is upset.* Nothing can top the observational skills of a trained law enforcement officer.

His banging provided sufficient incentive for me to look up and see a look in his eyes that caused concern. There was a wildness there that communicated a lack of fear about any consequences I could possibly bring to bear. Adversaries who perceive they have nothing to lose are the most unpredictable. I foolishly reiterated that I thought he was deserving of a citation, stoking the fire.

"Oh, yeah! See what you think of this!" he yelled in a voice much too loud for the circumstances. With that, he deftly used the end of his crutch to sweep my portfolio onto the floor. Still hoping my outward calm would have an infectious effect, it occurred to me that this crutch-wielding individual occupied the path between me and the door to the outside. As I reached down to the floor to pick up my portfolio, the man reached out and grabbed a generous handful of my abundant hair and pulled. Young and in pretty good shape, I quickly reached up and grabbed his wrist, first with one hand, then with both, to try to pull it free. When he did let go, he fell backward and hit his head on another one of the chairs around the table. As he hit the floor, his left pant leg slid up far enough to reveal a wooden leg, the scorched outside of which had apparently gotten too close to the fire on at least one occasion.

Oh, no, my mind flashed. *My first full fire season on the job, and I've killed an old man with a wooden leg.* But he stirred, and almost reflexively, I leaned over to try to help him. As I started to pull him up, he snapped off an impressive right that caught me full in the left cheekbone. *I guess he's okay,* and I quickly eased him back down to the floor. I grabbed my portfolio, announcing to whoever might hear that I would send a ticket in the mail, and bolted out the door.

Thankful that I had extricated myself from the situation, I powerwalked sixty feet to where I had moved my truck, aimed toward the road. *I'm glad that's over,* I thought as I began to regain my composure. I grabbed the radio, still shaky, to inform Louie of my situation. I began to speak when, to my horror, I saw to my left the man approaching the truck at a pace that betrayed both his age and his one-leggedness. He was waving his crutch and screaming at me. I'm not sure how I ended the communication on the radio, but I did so quickly. As I went through the endless process to start the truck (it was a diesel, requiring a few seconds for the glow plugs to warm up), the man reached the truck just as I hit the ignition. In his wild-eyed exasperation, he raised his crutch and smashed it over the cab of my truck, splinters of wood flying in all directions. I roared down the driveway and out onto the highway.

A quarter of a mile down the road, I did a U-turn and stopped on the shoulder to monitor the next potential assault from the old man. Still shaken, I called Louie again. When I finished whatever hysterical words I said to Louie, I still remember his calm, even response: "Spring Green ranger, do you require assistance?"

I don't know if Louie could sing, but at that moment, his speaking voice possessed the deep, rich smoothness that I imagined Bing Crosby's speaking voice would possess. It was a balm. I quickly rejected reflexive responses like, "Hell, yes," or, "Well, what do *you* think?" I soaked in the calm emanating from Louie's voice. Deep breath. "Ten-four." Louie was doing his job masterfully. "Requesting some law enforcement assistance."

"Should I ask them to meet you at the house?"

Let's get that straightened out right away. "Negative. I'm not going back to the house. I will meet him on the highway just up the road."

"Understood. KXT996." Later, I had an image of the growing awareness of a field problem (me) in the office at Dodgeville and a growing crowd of people gathering around Louie and the radio, monitoring the developing drama. With Louie's sign-off code (required by the FCC) at the end of the communication, I imagined a response similar to that elicited by police counsel at the scene of an incident: "Okay, folks, move along. Nothing to see here. We have things under control." That's what Louie did, how he impacted situations and people.

In a few minutes, I saw a county squad crest the hill in my rearview mirror. He pulled up behind me. I got out of the truck, and we spoke on the shoulder as I explained all that had happened.

"Do you want to go in there with me?"

Didn't you hear what I just explained? This man tried to kill me, and he almost destroyed my truck. You've got a gun. All I carry is a shovel. Do you think I want to go in there with you? "Maybe I'll pass on that."

"Okay. No problem. I know him a little bit."

I'll bet you do, along with every other deputy. He's crazy. "Okay," I agreed in feigned disinterest. *Thank you, Lord, that I don't have to go back there at risk of life and limb!*

I waited in the truck for him while he entered the compound. In a few minutes he came back out, reporting that the man was quite calm. Sometimes, it influences people when you're wearing a gun.

I returned to the ranger station, wrote the ticket, and sent it off in the mail, certified. As I recall, there was some problem with him choosing not to pay the citation or appear at the court date. I had to do some follow-up work in the court system, but there was no more direct contact related to this incident. I was also a little perturbed that there would be no action against him for punching me, probably a product of young guy versus old-man-with-a-wooden-leg reflection.

A few months later, I saw the man in town again. He had a big smile for me, of a sort that reflected that he knew he had pushed me over the edge at our previous meeting and the satisfaction associated with that. He was surprisingly friendly, even through my cautious engagement with him. I never had another problem with him. I did get some teasing from some of the guys with the fire department who heard my hysteria on the radio after the incident, as well as Louie's professional response. It was good-natured, and I had it coming.

It's Always Been Ours

The locals called it a marsh, but it was really a very high-quality lowland prairie, the best example of its kind east of the Mississippi, some said. Almost three thousand acres (more than four and a half square miles), it ran for several miles along the Wisconsin River. It was marginally separated from the mainland by a tiny creek that escaped the river at one end and spilled back into it at the other. This little creek was rarely more than a few feet wide, so it was only a minimal separator from the land that was part of the small village on the other side. The prairie served as a buffer between the town and the river. At its west end, a series of wooden power poles escorted an electric highline across the marsh to a point where it could reach across the main channel of the river.

The marsh commanded significant local interest as there were good hunting opportunities there for pheasants, waterfowl, and whitetail deer. As a result, there was a desire to keep the marsh open and free of any woody vegetation, an objective shared by those whose sole interest was in the quality of the prairie. Historically, there had been frequent fire activity, much of it clandestine, to set back potentially encroaching vegetation. On occasion, burning conditions and the direction of the wind had created a threat to the town, although local folks never seemed to take it very seriously. As with elsewhere in Wisconsin, springtime was the best time to get a fire going there. Investigation into who was responsible for such fires invariably led to a black box supported by shrugging shoulders. Despite the illegal nature of the burns and the occasional threat to the town, there wasn't much concern locally, and it always seemed to be a point of interest, even pride, when the authorities were defied. And the covert fires continued to occur, safe within the secret structure of the little town.

The stories of past fires filtered down to me from older rangers who had moved on to other locales. The marsh was largely inaccessible, so for the most part, the extent of the suppression effort was to monitor a fire as it burned along the creek that separated the marsh from town so that it did not cross and create a threat to the village and the homes there. Depending on the direction and strength of the winds, this could be a simple task, or it could be a challenging problem, continuing until all the grass along the creek had burned out. The only other concern of note was the power poles at the west end of the marsh. Again, depending on the wind, it could take a couple of days for the whole marsh to burn if it didn't get rained out. So, if it became clear that the fire was going to reach the poles eventually, then the dirty and difficult work of backfiring around the poles ahead of the main fire was essential to maintaining the local power grid. Burning around the poles was a relatively cool way to create a small break around them and prevent them from catching fire when the hotter, more aggressive main fire arrived. Backfiring the poles involved hip boots and a hike of almost half a mile to the furthest pole (there were six) and through water that could be as much as two feet deep, all over soft ground. The timing had to be right so as not to get caught out there when the main fire came. The whole operation was slow going because of the water's presence, which always struck me as ironic.

I made an effort to corral random local burning and to better coordinate the burns with appropriate weather conditions, both to more effectively kill encroaching vegetation and also better provide for the quality of the prairie. We spoke regularly with the local folks about burning plans so that those in positions of knowledge and authority were aware of our efforts.

One particular year, we were close to the time to burn, simply waiting for the right combination of humidity and wind to burn off a large chunk of the prairie. We generally enlisted the help of the local fire department for manpower since it was difficult to get equipment out there and also to include their official components in the whole burning effort. As a result, we were in ongoing contact with the fire department about our plans. We were burning later in spring than the locals historically had because it was more effective in setting back invading vegetation, much of it willow, to burn it after it had leafed out. But the delay bred some local impatience.

On this day, I spoke with one of the fire officers and told him that the humidities would be too low to conduct the burn safely that afternoon but that we were looking at a day later in the week as likely being appropriate. The information was received matter-of-factly, and I was content, for the time being, that we were on good footing. But later that afternoon, I had a sense, or a hunch, that something wasn't right. I called my ambulance crew chief to tell him I'd miss the meeting that evening and then called my boss to share my hunch. I suggested to him that we monitor traffic passing by the lone access road into the marsh in case my guess was correct. He agreed, and we met in Spring Green a little later to make a specific plan.

There is a highway that runs parallel to the river, through the town, and it is from this highway, near the most upriver end of the marsh, that the lone access road into the marsh departs. Rick and I positioned ourselves just off the main highway on opposite sides of the access road, significantly off the highway, so we'd be difficult to spot. There were no other side roads off of the highway between our positions, which were perhaps three-quarters of a mile apart. The road Rick was on went up a slight hill, so he not only had a view of the highway but also some perspective out into the marsh. And so we sat, Rick in a car and me in a nondescript (waterless) truck, speaking back and forth on a point-to-point frequency on our radios, one that anyone else would be highly unlikely to be monitoring with a scanner. We did not have a view of the road into the marsh, but any vehicle passing either of our positions would have to pass the others unless it either stopped in between, and there would be few reasons to do so, or it went into the marsh. A car would come by me, I would share some descriptive information about it with Rick, and a minute or so later, he would report it, passing his position, too. And vice versa. This was a long, tedious, and boring activity at the end of a long workday, so we maintained a running conversation about whatever crossed our minds in between handing off vehicles.

The day's light was fading, and we were beginning to entertain the idea that my hunch was a bust and that it was time to call it quits. We had been out there almost two hours when Rick, in the position nearer the little village, reported a vehicle with some unusual characteristics passing his position. I commented that the vehicle sounded like one driven by one of the local authorities and laughed at the absurd notion. Rick also observed that it wasn't going particularly fast and would take a while to get to me. We went back to whatever it was we were talking about, but after a while, Rick asked if that vehicle had passed my position. Huh. It had not, and I told him so. Our antennae were up. But as was frequently the case, we hadn't planned what to do if we actually encountered activity confirming my hunch. Over the next five to eight minutes, we confirmed that the noted

WHEN THE SMOKE CLEARS

vehicle had neither passed my position nor returned past Rick's. We began to discuss what to do, and we did so cautiously on the off chance that someone was monitoring our conversation.

As we continued to discuss how to proceed, Rick suddenly announced, "I see fire out in the marsh. I'm going in." In the adrenaline rush of anticipation, we both moved toward the access road. Rick arrived first and headed directly into the marsh. I stayed at the point on that road that crossed the little creek to ensure no other vehicle either came into or went out of the marsh.

The lone access road Rick used was situated on the upriver side of the marsh. As he headed in, he saw that there was significant fire to his right, farther upriver. There were only fifty or seventy acres in that direction before running into water, and the wind

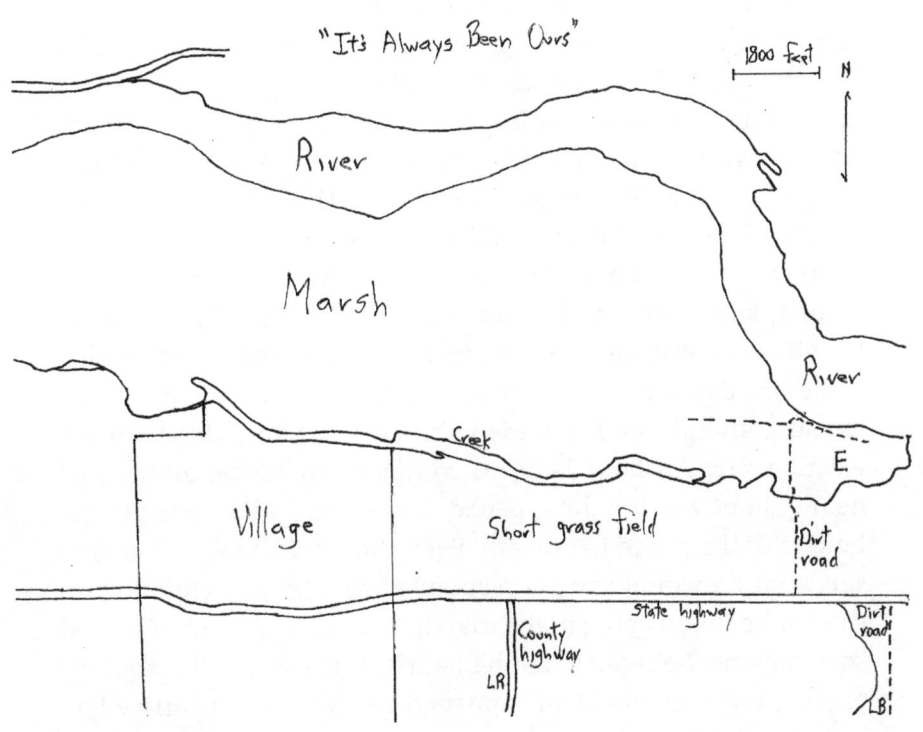

A sketch of the marsh area. The dotted line north of the highway marks the dirt road from the state highway into the marsh. Where that road crosses the waterway, Gramma's Oldsmobile wouldn't make it, but a high-clearance truck or SUV would. "LB" marks where I was positioned to monitor traffic going by the marsh road. The other ranger was at "LR." "E" is where the encounter with the arsonist took place. Sketch by Blair Anderson.

that night was blowing from that direction toward the bulk of the marsh to the west. Farther in, he could see there were places where the fire had already crossed the access road, meaning the fire was already beyond control and would burn through the night. Rick drove near the river, to where the fire had obviously originated. In another one hundred yards, he encountered the infamous truck we had seen go in the marsh. Soon after, he saw a figure through the smoke with a fusee, or road flare, lighting grass. Rick got out of his car and hesitantly approached through the charred ground. When some of the smoke cleared, he could see that it was Perry, the police officer from the local village, who was also part of the fire department. As the man turned toward Rick, Rick could see that Perry was in full police regalia, including a sidearm.

They looked at each other as each assessed the situation. It's difficult for me to imagine the thoughts that must've been going through each of their minds. Rick, of course, had to have some immediate concerns regarding his safety. And Perry, for his part, had to have an awareness wash over him that he, a police officer, had just been caught committing a felony. The realization of what the implications of that would be for his job, and perhaps for his freedom, must have caused him to consider, even for a split second, some extreme options. As I learned later, Rick handled the situation very casually, approaching Perry and calmly asking if the fire department had been notified. Smart. That broke the tension, and all the focus then shifted to what to do about the existing fire. An hour later, in containment mode, after I had heard about what had happened from Rick, I asked Perry what he was doing out in the marsh. Even though I knew, I wanted to see what he would say. He responded that he was checking on the fire he found going in the marsh. When I asked him if he had seen anyone, he replied that he hadn't. Perry was still struggling to piece together what had happened and how we had arrived on scene so quickly.

The rest of the night was spent with the logistics of trying to limit the impact of the fire, assuring that it didn't cross the creek separating it from town. A watch was arranged through the night, staffed by members of the fire department, sufficient once the winds calmed. It wasn't until the next day that discussion about the setting of the fire was broached in earnest. Perry was remarkably cooperative and didn't seem to be either fearful or defensive, perhaps realizing that circumstances had him in a place from which he could not extricate himself. Or maybe he genuinely didn't think torching the marsh was much of a problem; the impact of local culture can run deep. Over the next days, follow-up questions, many of them a repeat of previous questions, began to put holes in Perry's story that someone else had started the fire before he arrived. It wasn't until several days afterward that he asked how it was that Rick and I were there so soon. When I told him, I could visibly see his countenance fall as he sensed another insurmountable problem with his story. Eventually, he admitted, in a general sense, to his involvement in starting the fire.

But we needed something more specific, more substantial. To gain that, we arranged for Rick and the forestry law enforcement specialist from Madison to conduct a formal interview with him in Perry's police office to document the details of his story and get his formal agreement with it. The plan left me out of the process because, as the local ranger, I needed to maintain the best day-to-day working relationship possible with the local fire folks, and this interview was a key component in a process that was going to have a significant negative effect on one of their own. The local attitude, of course, was that starting the marsh on fire was not a big deal, regardless of who had done so. So there was a sense that Big Brother was coming after the poor little local boy.

The interview took place in Perry's office, again with him in full police regalia. Very specific questions were asked, and the responses were systematically written down. Unclear or nonspecific responses led to a relentless clarifying follow-up, which was wearing on Perry's effort to avoid the stark truth. When the interview was over, a clear picture was painted of Perry acting of his own volition, deliberately setting fire to the marsh. It was very damning. To this day, we don't know if others were involved. Perry never spoke to that. But it wouldn't have benefitted if he had. That information is forever in the community's secret repository, time perhaps gradually transforming it into legend.

The statement was read back to Perry, and as it was read, he was frequently asked if what had been read was accurate and if any mistakes had been made with what was written down. Interrogators deliberately incorporate and correct a few mistakes to demonstrate the process that took place to enhance the statement's veracity, especially for court proceedings, and that was done here. At the end of the reading, Perry was asked to initial each correction and the bottom of each page and write at the end that the statement was accurate, and then sign it. He was then encouraged to make and keep a copy of the statement. So he took the statement, left the room alone, and went elsewhere in the village offices and made a photocopy. On returning, he gave the original document to Rick, and the meeting was over.

In court, Perry's lawyer introduced several motions in an effort to suppress the confession. One was to have the suppression disregarded because Miranda rights (you have the right to remain silent ...) had not been shared with him. Miranda warnings, however, are only required for custodial interrogations. Custody, in the law, is defined as the perception by an average person that they are not free to leave the presence of the police agents. In the case of the interview that generated the confession, the interview took place in Perry's office. He had the only weapon in the room, and he got up and left, unaccompanied, to make photocopies. Beyond that, Perry, as a police officer, knew of his right to silence and to counsel. So we were able to easily defend the idea that the interview was just that, an interview, and in no way a custodial interrogation that would require Miranda warnings.

The other motion I remember saddened me. I knew Perry a little bit, and while we were not close friends, he was a man, and he had responsibilities. The motion was that Perry lacked the intellectual capacity to understand what was happening when the confession was generated. It was a desperate and demoralizing effort to quash the confession, and it was personally humiliating to Perry. Both motions were denied, leaving Perry and his lawyer without a leg to stand on.

The district attorney ended up agreeing to a plea bargain by which Perry would plead guilty to a misdemeanor charge, pay a fine, and resign his job as a police officer in the little village. Had he been found guilty of a felony, he would not have been allowed to own or possess a firearm. He was a police officer by training, and like most men who live in rural communities, he was a hunter. Even in my thinking, the fact that the offense involved the marsh was at least somewhat mitigating. I later learned that he was hired as a police officer in another community in a different part of Wisconsin. I assumed they knew about his conviction, and I wondered how they were able to justify hiring someone who had done what he'd done. For my part, I decided that it was not my responsibility to follow Perry around the state to ensure everyone knew everything about him. What he had done was wrong, to be sure, but within the context of the situation, I talked myself into some measure of understanding as to how he convinced himself to do what he did. And I thought of him not so much as a dangerous man who has no respect for the law but as someone caught up in a local cultural tradition who simply exercised exceedingly bad judgment.

Feigning Death

The problem of fire setting in the marsh abated for a couple of years, but not for good. After several springs with illegal fires, we decided to secretly surveil the marsh again. Since the locals now knew that we may be watching the marsh, we had to be more careful in how we set up. On one occasion, we put together a team of three for an evening when we thought there might be some activity. Only one had a vehicle, and they stayed in town where they could be easily seen. Another climbed the hill across the highway from the marsh where they could see the access road and if there was any traffic on it. As the third member of the team, I walked out the access road into the marsh and sought cover in case anyone came out there.

Out in the marsh, not far from the access road, there was a long shallow hole that had been dug to assess the depth of the water table years ago and had never been filled in. It was two or three feet deep, perhaps ten feet long and three feet wide, and struck me as something like a grave. I spent several hours crouched in there that night, undetectable but able to see for some distance around me. Between the inactivity and the cooling temperatures, I got cold. At one point, I thought I heard something, so I laid down in my grave so as not to be seen. As I listened, I heard movement and the sound of intermittent steps slowly getting closer and closer. It was then that I realized we didn't have a detailed plan about what to do if we actually encountered someone. I had a radio, but using that would give away my position. Surprise the source of the sound, or remain hidden for the sake of my own safety? If someone looked into the hole, what would I do? "May I help you?" seemed a little formal under the circumstances. I was actually somewhat fearful as my eyes strained to focus in the low light. Something long and narrow began to appear across the edge of the hole. The barrel of a weapon? It was ... an antler. As the young buck peered into the hole at me, we gazed at each other from four feet apart. We were similarly confused. I'm not sure who was more surprised. It was surely a first-time experience for us both, and neither knew what to do. My fear gave way to amusement. And I realized I had one of those entertaining stories, but it also didn't paint me in the brightest light. I decided then and there that I'd never share it with anyone else.

Did You See the News Tonight?

The recurring fires in Avoca had caught the attention of the media, and when we let it be known that we were going to do a controlled burn there one spring afternoon, a local TV network sent a team out to cover it. We had a large crew managing the burn, which was to take place in several separate areas, to ensure a good nearby seed from adjacent unburned areas to repopulate the annuals after the fire. Everybody had their assignments, and the weather cooperated, for the most part, so that we were able to accomplish our objectives with the burn. During the burn, I accompanied the correspondent who would be reporting live from the site after the burn was complete.

She was a young woman well prepared to be in front of the camera, dressed to the nines, but not ready to be out in the marsh. To her credit, she was fascinated with the whole process, peppering me with questions. My plan was for us to walk out the main road in the marsh and arrive at the far corner well before the fire backed to that point, when it would be uncomfortably warm along the road but not dangerous. As I assessed her attire, I was uncertain what might happen to her dress if it was exposed to too much heat. Would it burn or melt? Regardless, because of the timing of the ignition of the unit next to the road, we needed to make steady progress toward the corner at the other end.

She was curious about the burning strategy, the biology, and even the specific plants soon to be burned. I don't often run into people outside of our own biological circles who have much interest in these things, so it was encouraging for me to entertain her questions and help her understand, especially when I knew what she was learning might show up on the evening news and enlighten people.

As we spoke, I was visually monitoring the lighting sequence that was taking place on the far end of the unit along which we were walking. It became clear that we needed to pick up the pace a bit to be sure we got to the corner before the fire did. But this gal was tenacious in her thirst for knowledge, and I was surprised at the difficulty I had encouraging her to move along. At one point, when we were still less than halfway to the corner, I received word from one of my folks out farther in the marsh that there seemed to be a slight shift in the direction of the winds, which, if it continued, could turn the backfire we were observing into a head fire that would burn more directly into the fuel, generating more flames and significantly greater heat.

At this point, somewhat concerned, I took my reporter friend by the elbow and encouraged her and her high heels down the road. Soon the clear skies we had enjoyed directly above us began to fill with smoke. The wind had, in fact, shifted, and we were not in a good place. There was a ditch on the opposite side of the road from the fire, and there was still water in it, so I had no concern that the danger that may be headed our way was in any way life-threatening. At the change of circumstances above and around us, this reporter recognized the urgency of the moment and began moving along more quickly, with palpable concern. Soon, she kicked off her heels and picked them up so she could move more efficiently, making quite a mess of the nylons she was wearing.

But when the wind continued to shift in our direction, we were too late. Behind us, the fire was reaching the road, and twelve-foot flames angled across the road at a forty-five-degree angle. When we saw this, she became genuinely frightened, and our trot accelerated. Her questions had evaporated, and she was looking to me for direction. We still had seventy-five yards to go, and I knew we weren't going to make it. As the heat built, I reassured her and escorted her, willingly, into the heaviest, wettest, green vegetation I could find in the ditch. While I was wearing flame-resistant clothing, she was in a dress made of a material that I feared might melt in the heat, with the most unpleasant consequences. So we got down low in the ditch, with six inches of water, and I covered her in an effort to shield her from the heat, which was now building significantly. Soon the head fire reached the road across from where we were. It was loud, windy, hot, and I am sure, very frightening for her. Even through my Nomex clothing, my back was uncomfortably warm.

Just as quickly, it was over. Immediately, I helped this bewildered woman up, and we got back on the road, which had sufficiently cooled in the continuing breeze and now had smoke and ash blowing across it. We watched the rest of the fire burn out between us in the corner. After that was done, we looked at each other. We were both a mess, but only I was designed for it. She was dazed, wet, and dirty with vegetation. Her hair was disheveled, her nylons were a mess, and we were only able to find one of her shoes. She actually thanked me for saving her, and I assured her that the danger had not been as great as it may have seemed. I apologized for putting her in that position and treating her the way I did when the fire reached us. She was very gracious in taking the blame for our predicament. As we talked, the ash stuck to her face and her wet clothing, further deteriorating her appearance. The final humiliation was a dog that appeared on the burn site. When my new friend and I encountered him, he had been delightedly running around and was absolutely filthy. He thought the reporter was the most interesting thing he'd seen all day and proceeded to jump up on her with his dirty paws and behave in inappropriate ways that male dogs sometimes do. I tried to defend her dignity, but there wasn't much left.

After a few minutes, we reunited with her photographer, who had been out filming some of the burn activities. When he saw her, he could not contain his amusement at her appearance. One more humiliation. I escorted them back to the creek, crossing to where their vehicle was, and said goodbye, imagining their conversation on the way back to the station.

We were busy with the burn and equipment cleanup until after the news had aired at six o'clock. What was shared about the burn was all film and voiceover, no visual of my reporter friend. I was the butt of wisecracks from my colleagues for quite a while.

WHO DO YOU TRUST?

In law enforcement, some cases end up in the courts. Officers are called to bring a clear historical account of particular events involving particular people and their actions to the district attorney, assembling a clear case that can be won in front of a judge and jury: "Who done it?"

With fire, especially wildfire, most people envision it only at its inferno stage. They have to be brought to the realization that virtually all fires start very small. It is then, in its infancy, that the cause can most easily be determined. Most forest rangers can weave the truth together pretty effectively to convince those who need convincing, but unlike TV, there is frequently at least a small shadow of doubt in each decision maker's mind, be it the district attorney, the judge, or the members of the jury. Some of those doubts are more easily overcome than others.

The best evidence is when the person responsible for the fire, the one who made the mistake, freely admits the mistake and that they took an unwise action that resulted in starting the fire. A confession is hard to beat.

But by our nature, we don't want to admit a mistake, especially when we know it will cost us money. Most are honest, but a few of those responsible will fabricate an erroneous version of history, of how something happened, that has nothing to do with them, and they try to get you to believe it. And when circumstances clearly tell you their version is not the truth, that sticks in your craw.

Such liars are generally not easily dissuaded from their efforts. It can turn into a wrestling match. Tentative challenges like, "Are you sure you didn't burn anything in that barrel by where the fire started?" are easily rebuffed because of the questioner's uncertainty. Some liars get on a roll with their story and gather momentum. The more fantastic the story gets, the more their confidence grows, and the harder they are to reel in.

Liar

Early in my career, I dealt with liars indirectly, with limited success. Sometimes, I wrote up my incident report based on what I believed happened, regardless of the landowner's denial, issued a ticket and sent them a suppression bill. That worked sometimes; the prospect of repeating the untruth in front of a judge eroded resistance, and the matter resolved itself. But sometimes that approach didn't work. And again, that stuck in my craw.

I had to figure out a way to challenge liars more directly. But it's hard for most of us to generate the nerve, the chutzpah, to tell someone, face-to-face, that they are lying. In casual conversation, that rarely happens, and when it does, it is a character-related personal challenge, generating considerable discomfort. Still, it was part of my job to overcome untruth.

Sometimes fatigue, like alcohol, can be a great liberator of the tongue, usually generating problems. But one Sunday afternoon, on my sixth fire of the day, the fourteenth fire of the weekend, fatigue was my friend. I arrived in a fire situation that the local fire department had already controlled. It was a textbook case; the origin and cause were obvious even before my truck came to a halt. Before me was a typical V-shaped fire with a burning barrel at the base of the "V."

"Up by the house, in the blue shirt," the chief smirked without a greeting. "Good luck."

"I wonder what started this one," I replied, facetiously.

Blue Shirt was holding forth on the open porch of the house in bellicose fashion with his posse of family and friends. He was a bigger fellow, disheveled enough to convey only a loose command over his circumstances. With a mustache about ten years past its time, the affect was confidence, but the apparent need for an audience betrayed a latent insecurity.

"Excuse me." The request form was designed to suggest a notion of subservience and give his sense of control a boost, which can be helpful when trying to get someone to admit to something not in their own best interest.

He turned toward me, beer can in hand, glanced at my badge, and turned back toward his crew with a broad, self-satisfied grin. "Uh-oh. I'm in trouble now."

I thought to myself, *That worked well.*

"Can you tell me how the fire started?" The show had begun.

Act 1. "Well, we don't know for sure. We were all in the house, and somebody looked out the window and saw the fire, so we called 911." He glanced back at his posse. "There were some neighbors riding around on four-wheelers earlier. I suppose that must've been what caused it." One or two heads on the porch nodded almost indistinguishably. But he didn't notice, one way or the other. He was in performance mode.

Act 2. Stories of the neighbor's irresponsible actions grew as the diatribe continued. Liars, particularly when liberated by a little alcohol, as Blue Shirt seemed to be, tend to tell increasingly fabulous stories with more and more detail, especially when they have their audience with them. The more outrageous the stories became, the more in control he felt and the more oblivious he became to his surroundings. But he was overwhelming me with his performance. In my growing frustration, I felt my face flush, and the muscles in my neck tightened. As he continued in his own world, something snapped in me.

Act 3. I took a step toward him, now only a couple of feet in front of me. My glare narrowed, and I looked at him sharply, intensely in the eye. *That* he noticed. In the middle of his sentence, my words almost exploded out of my mouth. "Stop! You and I both know that what you're telling me isn't true." In the pause that followed, you could hear a pin drop as the force of my expression seized everyone present, me included. For a brief moment, I was wordless. I was trembling a little, but I sensed I had turned the table. "Now why don't you tell me what really happened out there."

And then, after a silence that seemed to last an eternity ... his head went down. "You're right," barely above a whisper. At least one person on the porch turned away to conceal a grin. I was stunned, so much so that for a moment, I wasn't sure what to say next. I had faced down this showman and won. I looked to the porch, and almost everyone immediately began milling about, looking for something, anything else to occupy their attention. Blue Shirt was most cooperative during the ensuing interview.

That may have been the single most confidence-building event in all the years of my enforcement work. After that, I learned how to contain my emotions, and such confrontations became a most useful tool in the arsenal. At some level, almost everyone wants to experience the freedom of the truth, and under certain circumstances, they just need a little help getting there.

ORIGIN AND CAUSE

In most fire circles, the phrase used to figure out how a fire started is called "C&O" for cause and origin investigation. But since one generally needs to find the origin to determine the cause, I think of it the other way around.

The key to resolving the story behind a fire is knowing what caused it. And the best insight into that is looking at evidence in the area of origin where the fire began. Most of the time, especially on smaller fires, less than an acre, the origin and usually the cause are as plain as the proverbial nose on one's face. Where fuels and topography are consistent, and there is a light breeze, the fire burns away from the origin in a "U" or "V" shape, burning downrange faster than laterally, in proportion to the wind; the stronger it is, the longer and narrower the alphabet. Somewhere near the base of the "U" is where it started. And if there is a burning barrel or a campfire or a blazing brush pile there, any of us can figure out what happened. All that remains is the who, and that's usually the guy with the look of relief on his red face who's not dressed for the occasion. Regardless, the area of the origin is to be suppressed gently and protected vigorously.

Fires along the road were almost universally attributed to cigarettes.[11] But that was too easy an answer, and I didn't believe they were a capable ignition source. Back then, the side of the road was strewn with cigarette butts, most of which were cast out live. If cigarettes were as prolific at starting fires as "they" said they were, the whole countryside would have gone up in flames.

Most roadside fires are started by some sort of equipment. A red-hot piece of a deteriorating catalytic converter may break off and hit grass fuels along the road. A farm tractor, exercising its throttle for the first time in spring, may dislodge a piece of hot carbon from its exhaust system, especially as they are "revved up" to scale a hill. One time, a fellow had burned a brush pile and, thinking it was out, loaded the ashes into his farm wagon behind his tractor to haul to a different location to dispose of them. Unfortunately, there were still hot embers in his pile. They fanned to life as he drove down the road and blew out. I recollected that there were nine separate fires along a three-quarter-mile stretch of road before he realized what was happening.

One of the things I always stressed with the fire departments in training was protecting the area of the origin where the fire started. Key clues and evidence can be found there, which can point to the cause and who may be responsible for starting the fire. Protecting the origin takes some fast and thoughtful appraisal when first arriving on the fire. Usually, the fire has burned itself out at the point of origin by the time suppression resources arrive. A couple of departments took that protection role so seriously that they assigned one of their guys to locate that general area and protect the origin from fire vehicles, water damage, dragged fire hoses, and curious onlookers. I appreciated that.

---

11 I had become quite dissatisfied that many fire people were quickly ready to attribute almost any fire on any road as having been caused by smoking. Back in the eighties, roadsides were littered with cigarette butts. If they were as adept at starting fires as everyone liked to think, the whole state would have burned. One dry spring day, I bought a pack of cigarettes and tried to start a fire with all sorts of fuels and every sort of assist I could think of, including nesting the cigarette in a bunch of grass and blowing on it for a prolonged period. I was never able to start a fire and concluded that roadside fires needed to be investigated more thoughtfully.

Dear John

"It is better to have loved and lost, than never to have loved at all."
—Alfred Lord Tennyson

Once in a while, there was a fire that was perplexing upon initial inspection. One fine spring day, I rolled up on such a fire. It was in the middle of an unplowed cornfield, with heavy corn stubble from the previous fall. So no farm equipment had been there since fall. The only vehicle tracks were from the lone fire truck that quickly squirted it out long before it reached the edge of the field.

As I approached the chief, whose resources had beaten me there by ten minutes, he greeted my perplexed expression with a smile.

"You must know something I don't know."

"I do indeed," Mike smiled. He was a sharp guy, a good chief with a disciplined department. I always enjoyed working with them and was very confident in their capabilities and their willingness to work cooperatively. Mike, who paid his bills as a photographer, took fire pretty seriously. So when he was as breezy as he was, I knew something unique was up. "Let's go to the barn."

The farm buildings from which this field and others were worked were around the bend of the road. I jumped into his vehicle, and we went to the barn, where a small group was gathered, apparently in some sort of mourning.

Knowing my penchant for strong law enforcement, Mike counseled me as we started out of the car. "Take it easy on this one." I respected his counsel, though it only made me more puzzled. As we walked up to the cluster of people, Mike's eyes met the man in the group and nodded toward me. Seeing the questioning look on my face, he released the care of a sobbing girl, maybe fourteen, obviously his daughter, to her mother and approached me.

"Terry had this boyfriend, her first, really, nothing serious," he started right in. Seeing no alleviation in my quizzical expression, he continued. "It didn't work out." I started to look around for

Geraldo and the cameras, but I was not getting the connection. Finally, explaining it to me as though I were an incompetent, he said, "She got a letter today. You know. Like Dear John." My wheels started to turn productively at last, and I was beginning to track.

"Oooohh," I epiphanied, much to his relief. "Out in the field," and his head started nodding.

"She's pretty upset about everything."

Having been deceived before, I always wanted things firsthand. "I'll go easy," I assured him as I approached the family cluster. The poor girl was distraught enough about the romance, and to have to make it so public was humiliating.

"Hi, Terry. I'm sorry," I opened. "You burned the letter in the middle of the field?" She nodded, and the tears reenergized. Enough. I put a hand on her shoulder. "I'm just glad that the fire was so small." She nodded and refocused on her misery, safe in her mother's arms.

One of the maxims of law enforcement is that its purpose is to achieve compliance with the law. With some folks, that takes an enforcement action, like a fine or something more severe. But not here.

My work here is complete, I thought. And it was.

Lesson Learned

"I will discipline you but only in due measure."
—Jeremiah 30:11

I have observed that the objective of law enforcement is to get people to behave or to influence a change in behavior that prevents the recurrence of illegal behavior and its damages and risks. Judgment is required to identify the action that most effectively accomplishes that at the least cost/price. Sometimes, it is challenging to pinpoint that. Sometimes it's not.

A small fire in a remote part of the county had been quickly suppressed. Before I visited, the chief told me that only one unit had actually gotten to the fire (all that was needed) and that, unusual for this department, they would not be billing us for their services.[12] A department's inconsistencies in billing can be problematic, resulting in the perception that the "out" people get billed and the "in" people don't. But I decided to wait to express my concerns with the chief, whose judgment I trusted, until I was more familiar with the situation.

Any gravel that may have once been part of the driveway had long since yielded to dirt. There was a small barn, so long past the need for paint that one couldn't be sure it had ever been painted. It looked tired but was sound, and the roof appeared to be effective, at least on the side I could see. I immediately observed a very diminutive fire that had escaped a burning barrel on the opposite side of the house from the barn. I knocked on the door.

When a woman opened the door, she gave a kind but regretful smile and let out a long, deep sigh. It seemed born of an eternity of burdens of which I could only speculate, much more than should be generated by one so young or so diminutive. Everything was just a little weary—the clothes, the door, the tile on the kitchen floor, and the look in her eyes. Her clothes were clean, and she looked like she had taken the best care of herself that she could. And the same was true of her young children. But all was as worn as the driveway and the barn. I was reminded of pictures I had seen of depleted landowners in the Dust Bowl. Her two small children appeared a little behind her at the door, one on either side. They were confidently curious about my presence but quiet and respectful.

"Hi." I briefly explained who I was.

"Yes. Bob told me to expect you." Her kindness maintained the upper hand on anxiety. "How much will this cost?" she asked, plaintively. It was the question that had obviously been playing through her mind since the fire, the one that would most immediately affect her small tribe. What sacrifice would have to be made to right her moment of poor judgment, a desperate process of prioritization that yielded only painful solutions? I

---

[12] Since we, in turn, billed careless responsible parties for the fire department's services, there would be no bill for the landowner, an incentive that can be a useful compliance tool. I still had the option of issuing a citation (ticket).

sensed a resolve on her part to pay the cost because that was the procedurally appropriate thing to do—the good example for her young children. But I knew that to do so would also impact this family in a way I had never experienced.

"Before we get to that, please tell me what happened."

"I set fire to a few things, and I knew right away I shouldn't have. It was an hour before my burning permit time, but the kids were hungry, and there were extra chores last night, and I just wanted to get it over with. I know I shouldn't have done that."

I felt an affection for this woman's integrity and for her little family. She was openly truthful, even when she couldn't afford to be, an even better lesson for the kids. I knew I would not have to do anything to avoid further trouble with her. That realization generated in me a smile that must have seemed inconsistent with the presumed purpose of my visit. "Then you surely know not to do that again," I confirmed through the smile.

"Oh, no." I knew she meant *yes.*

"Okay. Here's my card if you have any questions. Otherwise, have a good afternoon."

As she realized we were done, that there would be no bill, the tension that had held her shoulders so stiffly dissolved, and the lines in her forehead and around her eyes lost a little of their depth. Grace, like forgiveness, is one of those things that's great to receive but even better to give, especially when it is appreciated. As I left, I mused about the conversation she would have with her children and how her kindness and resolve would reinforce themselves in the perspective of her children. And who knows how they might express that in the years ahead.

Right Before My Eyes

Fire is a fascinating adversary. But it's the people that make things really interesting. And when you can mess with them, just a little, it can be fun.

Driving down the highway on a thus far quiet spring fire afternoon, I came upon a man burning a modest brush pile. It was not a problem, nor would it have been, if it had escaped into the narrow ditch of vegetation adjacent to it. There was limited fuel, and it couldn't have gotten far. But I had not approved a daytime permit there, and it was bad business to have unauthorized daytime burning going on, especially along a well-traveled highway. So I pulled in.

The man tending the fire sauntered in my direction. "What do you want?" he asked, belligerently.

Alright, if that's the way you want it. I cheerfully asked him if he had a burning permit.

"No. I don't need a permit for a little pile of sticks."

I responded with my regular schtick, still cheerful. "In spring, a permit is needed for any outdoor fire that's not a campfire. Do you have a permit?" I knew he didn't.

"That's ridiculous."

His back was to the fire, and a feisty gust of wind lifted a burning leaf from the pile into the adjacent ditch. Over his shoulder, I watched it gradually ignite a couple of adjacent leaves in the ditch, and it ever so slowly began to spread.

I explained to him how even small piles of debris, like his, could escape and start other unintended materials on fire.

"That almost never happens," he corrected me with an air of authority.

"Actually, it happens more often than you might think. In fact, you can see an example of that right next to your own brush pile."

With that, he turned around and saw that the fire had, in fact, spread into the ditch and was slowly growing. When he turned back to me, he had a different look on his face and a different attitude.

"When something like that happens," I gestured toward the fire, maintaining an almost academic air, "adjacent materials can start to burn, and if it is not suppressed, it can turn into quite a mess. When we issue burning permits, we look at the site and give people suggestions for how to avoid just this sort of thing."

It was hard to tell if he was more frightened by the fire that was escaping or by the fact that I just continued to talk without reacting with the same panic he was feeling. I was enjoying this a little more than I should have.

"In this case, I might have suggested that you burn your pile of sticks a little farther from the fuels in the ditch and perhaps have a hose nearby." My arms were folded, content with the ongoing discussion. I paused. "Do you have a hose nearby?"

He was morphing from frightened to angry and finally could not contain himself anymore. "Okay! Aren't you going to do something!?" he blurted.

"Sure, I can put the fire out." I still didn't budge. "I just wanted to make sure that you understood why we require permits and how dangerous burning can be," and gave him a moment to respond.

"I get it! Please put it out." I turned toward the truck. "Should I call the fire department?" he asked, frantically.

In one last impertinent power move, I stopped and turned back to him. "No. That won't be necessary." And then I added a little irrelevant information, just for effect. "I can call them directly, with my radio, if I need them." Pause. "But I won't."

I calmly walked back to my truck, drove slowly across the yard to the fire (so as not to damage the grass, of course), and in a matter of fifteen or twenty seconds, extinguished the escaped fire.

I returned to my new client. "So, do we understand each other?" It sounded a little snotty, even to me.

But he was relieved and submissive, emotionally tapped out. "Yes, I understand." He seemed genuinely repentant. I was feeling that way, too, and I yielded to a sense of grace that came over me.

"I'll just issue you a warning for this today, knowing that you will get permits in the future. But there is the matter of a suppression bill." His expression darkened. "Since I was driving by, anyway, there won't be any mileage charge. And the cost for my time here is, oh, about five bucks. Since we don't bill anything under twenty dollars, we'll just call it a wash. Okay?"

He was redeemed. We were friends, sort of. It ended well for both of us. He wouldn't be a problem again. And I had rather enjoyed myself and still managed to impart a little grace. All in about fifteen minutes. A good stop.

A-One and A-Two

"While the future's there for anyone to change, still you know it seems, it would be easier sometimes to change the past."
—Jackson Browne, "Fountain of Sorrow"

After ten or twelve years in Spring Green, I apparently acquired a reputation as a reasonably competent ranger. A class of new foresters had been hired, and the practice had been instituted to have them spend several weeks out in the field, some weeks with a more seasoned ranger and some with a forester, to gain a better understanding of different aspects of the program.

One spring, one of the new hires was assigned to Spring Green for three weeks. She was an enthusiastic young woman, excited to begin her career and make a mark on the resource. While her interest was probably more toward forestry than fire, the policy in those days was to create maximum flexibility among the forestry corps so that, in theory, people could move back and forth between fire and forestry, and forestry folks could provide better support to the fire folks in bad fire years. I was happy to help a new employee acclimate and learn. I almost felt like I was in a fatherly role with Shelley. She stayed in a motel in town, but she came to our house for dinner a time or two and met Jan and the boys. The time spent with the family made her experience in Spring Green more pleasant. When she came, it was spring, and we were fully into fire season, so I had a great opportunity to tell her about and experience many aspects of the fire program.

One time, we stopped a grass fire approaching a two-track road out in a field. It was actually a fairly dangerous tactic, but I had her drive, which put her on the side of the truck away from the oncoming fire, and I manned a hose behind the truck as she drove along, knocking down the fire as it reached the road. I would not have wanted to take such a direct approach on a fire that was any hotter than this one was, and it may not have been that wise to do what we did here. But we were successful, the truck wasn't damaged, and I wasn't hurt.

Another time, we were approaching Sauk City in the afternoon when we spotted a fellow burning a brush pile. It wasn't a particularly dangerous situation, as it was surrounded by lawn, and the nearest fuels were fifty feet away. But since I had to approve all daytime burning permits, and I hadn't approved one here, I knew he didn't have a permit. Even though it wasn't dangerous, it's bad policy to have people doing things they shouldn't, which in other circumstances could be significantly more dangerous. So we swung in, and I looked forward to the opportunity to show this young gal how to make such a contact.

I demonstrated proper procedure by calling in on the radio to advise our dispatcher where we were and what the situation was. We then exited the truck and approached the fellow who was tending the fire. I introduced myself as a local ranger, displayed my badge, and asked the man if he had a burning permit. "Nope. You don't need one for this kind of burning," he lectured me. I gently and politely corrected him, going through my routine about what the laws were and why they were in place. When I was done, he realized that I knew more than he thought I did and that I was going to control the interaction. He apologized and described the safe features of what he was doing, with which I agreed. I told him I would not issue him a citation because he was doing a good job with his burn and that he could finish burning what little was left in his pile but not add to it and extinguish the pile when he was done burning. I told him how and where to get a burning permit and reminded him that it could not take place until after six o'clock. I issued him a written warning, which we do mostly to get people into our system so that we know when they've been contacted. After this exemplary contact, we left, and Shelley and I discussed the contact in the truck. It was a good learning opportunity.

Two days later, we were on the same road, heading toward the Sauk City Fire Department to get some information about a fire run they had made the previous day. Out of the blue, Shelley laughed and covered her mouth. I looked over at her, and she looked at me. "What?" I asked, and she just looked back over her shoulder. The same fellow was doing the same thing in the same place at about the same time as two days previously. She was amused because she knew I wouldn't be.

Exasperated, I turned the truck around, called Louie, and described the circumstance and location to him. Louie knew me well enough to know from my voice over the radio that I was not pleased. In fact, he asked if it was the same location "as earlier in the week." (Louie was always the very soul of discretion.)

As we pulled into the driveway, a woman hurriedly came out from the house, apologizing and saying she would just pay the fine. She had his wallet with his driver's license and asked that I not bother him. I thanked her but informed her that I would need to contact him directly. As we approached the brush pile, yet again, he looked up and saw us and went into a profanity-laced tirade. I'm not sure if he thought we just sat in the office most days, but it seemed he thought the odds of us driving by his place when he was burning twice in three days were almost zero. In a tone different from the previous encounter, I told him that he would be receiving a citation for illegal burning and to get a hose and put out the fire. He resisted. So I also told him that if he did not want to put out the fire himself, I would bring my truck across his lawn and do it for him, but that that would generate a suppression bill for which he would be liable. He was fit to be tied, but he also knew there was no way out. I asked him why he burned again after our previous contact, and he angrily reiterated the safe features of the situation. I agreed but reiterated that it was illegal, and I had explained to him how to get a permit and burn legally after six o'clock. He was still upset, but he quieted and acquiesced to putting out the fire. With that, I told him we would get the information for the citation from his wife and that I would leave it with her. He muttered something and went off toward the hose by the house.

The poor wife was still standing by the garage, having witnessed the whole thing. When I got there, she had his driver's license out. I was sufficiently concerned with this guy's temper that I did not want to wait around too long, so I wrote down his driver's license number, enough to get the rest of what I'd need later, and we got out of there.

Shelley continued to be amused by the whole situation as we drove away.

Who's At Fault?

The spring right after I became Chief Ranger, I got more deeply involved in a particular fire than I probably should have.

At almost six hundred acres, the Crystal Lake Fire took out a house, some outbuildings, a number of camper trailers, and a lot of very well-managed and valuable red pine plantations. Although dozens of buildings and campers were saved, damages were significant. The Area Ranger there was as new to his job as I was to mine, and he was having a very busy first spring as a supervisor. He asked me, knowing my experience in law enforcement, to oversee the investigation of this fire. While that was his job and not mine, I did have a lot of experience in doing that, and my boss left the decision to me. It was a good opportunity for me to demonstrate that the folks in Madison, sometimes accused in the past of being detached from the field, were now interested in being involved in and supportive of what happens in the field. So, I hesitantly agreed to do the investigation.

The origin of the fire was fairly apparent. At the most upwind side of the fire was a very large pile of mostly burned red pine stumps, sixteen to eighteen inches across. They had been grubbed out of the ground, likely with a power shovel, so they included in their mass not just the stump of the tree but also a very large root mass, sometimes still encompassing dirt from between the roots. This pile must have included thirty to fifty of those stumps.

This fire took place in sandy country, so the ground fuels were a little thin. Because of that, most of the fire's advancement into new fuels was not because of lateral or backward burning but because of wind driving the fire forward. When the fire reached the pines, it crowned in several places; the fuels were *that* dry and the wind *that* strong.

One problem I faced in conducting the investigation was that I had commitments on the two days following the fire. The evening of the fire, I was briefly able to look at the fire scene when there was still a lot of mop-up activity going on and took some quick photos around the area of the origin. But it was a little haphazard because I didn't have much time or much light. When I returned to the fire a couple of days later, it had rained. In those lighter fuels, the rinsing of the rain made the indicators that were so clear the day of the fire much less apparent. Nonetheless, I took a set of forensic photos to demonstrate what had happened, even though those indicators were much more difficult to see at this point than they had been previously. I knew at the time that could be problematic.

One of the issues that made this fire interesting became apparent as I started asking some questions. The origin of the fire was part of a commercial property that included a small, privately owned lake. It was a heavily used recreational area where campers rented closely spaced sites to park their trailers and enjoy the lake and other activities available for campers. The campus included a playground, an indoor recreational area, and a store where various provisions could be purchased. Many of the patrons spent most of the summer there.

One of the first people I spoke with was the owner of the campground. He was a decent guy who immediately took responsibility for the fire. While I appreciated his willingness, he seemed almost too eager. When I pressed the issue, he told me that somebody "who worked for" him had actually set the brush pile on fire but that it had occurred under his "general" direction and at a time when the ground was snow covered.

I needed to know the specifics. "So, you told him to light the brush pile on fire that day?" I probed.

"Well, not specifically. But that pile was a problem, and we had talked about burning it as a way to get rid of it." He was a stand-up guy trying to protect someone.

"Okay. But you didn't specifically direct him to light the pile on fire."

"No, I didn't."

"Who decided to light the pile on fire?"

"He did."

"When did you know the pile was burning?"

"Actually, it wasn't until after the big snowfall that I noticed smoke, or steam, coming from the pile," he offered.

"Did you go out and look at it then?"

"No, but I asked him about it, and he told me he had lit the pile on fire a few days earlier."

"Okay. That's helpful. Thanks."

"If there's a ticket or something, write it to me."

"Well, I can't necessarily do that. But the court doesn't care who pays it."

"Okay. Let me know what you decide to do."

"I will. I might just write him a citation for failing to extinguish a fire. That'll cost about one hundred fifty bucks."

"Okay."

"But I'll let you know."

I appreciated this guy. When a lot of folks would have run for cover, he was doing his best to protect his unfortunate friend.

Further questioning of others revealed that the pile had been set on fire in February, a day or two before a major snowstorm was expected. The thinking was that the fire couldn't escape because the ground was covered with snow and that there was much more snow coming, making it an ideal time to burn up the material in the pile. It was sound reasoning, and I appreciated the logic in the approach. A check of weather records pinpointed the date it was set.

But here's the thing. With very large fuels like the stumps, especially when they're piled, and especially when there is a lot of dirt in amongst them, they don't burn efficiently, and fire can continue smoldering for a very long time, gradually drying out those roots still encased in soil in the process. Over a period of days and weeks, as the soil dries, it tends to fall away from the roots, exposing new fuel. It was a full eight weeks since the brush pile was set on fire that the wildfire escaped, and some questioned if a fire could persist that long. But I had seen it in the past and was convinced that that's what happened in this situation. In fact, a week after the wildfire, we were able to find hot spots deep in the pile.

As I looked further into the connections here, I learned that the individual who had set the fire had some sort of relationship with the campground owner. It turned out they were friends from a long time ago, and the one fellow had made some poor life choices, resulting in him being "down on his luck." In an effort to be helpful, the campground owner let this fellow keep his camper at the lake in summer in exchange for helping with various responsibilities at the campground. Mowing was one of those responsibilities, and the suggestion was made that perhaps the mower hit a rock, setting off a spark that started the fire. While that was conceivable, there wasn't any evidence that mowing activity took place in the area of the fire on the day of the fire.

I continued to believe that the blaze started from the brush pile and that the person responsible was the helper who had ignited the pile back in February. Had it been a lesser fire, I probably would have dropped the whole thing, but I thought there might be further reverberations because of the damage, so I took the most limited action I could, short of doing nothing. I wrote this fellow a citation for failure to extinguish a fire. It was an accurate charge and was, in my opinion, the reason we were all having this discussion. From a statutory standpoint, that seemed to be the end of it. The man who started the fire was virtually indigent, so he was not going to be able to pay the large fire bill that would result from suppressing the fire, let alone all the damages. Since you can't get blood from a stone, I figured the matter was closed and that anything that needed to be done in the future could be left to our local folks. I later learned that the ticket was, in fact, paid by the campground owner. Good man.

I always took pretty good notes in my investigations, and my incident reports, while simply factual, went into a great amount of detail, what seemed to many like more than was necessary. One of my bosses used to laugh at how I always included a statement that I either "displayed my badge" or that "my badge was displayed on my uniform." I did so because our authority to do a number of things at a fire scene is based on the fact that we are law enforcement officers, and people cannot interfere with those activities when they are aware of that. Making the statement about displaying my badge makes it clear that everyone present knows what I am. So it has some tangible benefit, should matters go to court, even though my boss found it worth a chuckle.

In the case of the Crystal Lake Fire, my anal approach to report writing would cut both ways. It explained in detail what I did, but it also showed that my investigation was delayed by a couple of days, a potentially damaging detail. It was all there for everyone to see.

It wasn't until a couple of years later that things really got interesting. The civil side of the case, related to the damages and involving numerous insurance companies, had been working through its process in the courts. When insurance companies are involved in such matters, as they almost always are when covered losses pile up, the value of personal relationships tends to take a backseat. Insurance companies insist that the covered individual sue to recover the loss, or the insurance company won't cover it.

Here, there were some very significant damage claims resulting from the lost structures and campers and the burned timber. The insurance companies for the plaintiffs, who had filed claims for their losses, were looking for another source of liability to cover those costs. The man who started the fire was not that source; he had no insurance coverage and no resources of his own. So the insurance companies were sniffing higher up the food chain.

There were a couple of potential alternatives. The first was that the fire was started by a source other than the brush pile, a source which would point back more directly to the campground owner, who, in this scenario, had the deep pockets. The second alternative, more nuanced, legally questioned the independent status of the man who started the fire. That is, was he acting on his own, in which case they were out of luck, or, although he was not formally "employed" by the campground owner, was there an agency relationship between the two of them? In English, even though the campground owner didn't specifically tell the man to light the brush pile, did the man light it in the general sense of fulfilling his "helper" role at the campground, one for which the campground owner compensated him by letting him use a campsite at the lake for no charge?

Over the past two to three years, the insurance companies, both for the plaintiffs and the campground owner, had wrestled with each other, had not come to terms, and had ultimately filed civil lawsuits related to the damages. After all that time, those suits had been scheduled for trial, and in anticipation, depositions were being taken to gather further information to present at trial. Guess whose name came up?

This was new ground for the DNR fire program. While our staff attorneys had had some experience with litigation, it was related to regulations and not to civil matters with big bucks involved. Because the question of how I did my job could potentially come into question, I was assigned counsel from the Department of Justice to help represent me through these proceedings. Everybody needs a lawyer.

As I recall, we had several weeks of lead time before the deposition, so I had the opportunity to spend some time with my attorney going over the case. She was young, a very short woman of Southeast Asian origin, and very quiet. Initially, I feared her quietness conveyed a lack of understanding of what I was sharing. But I was wrong. She turned out to be sharp as a tack.

Since I was the investigator on the case, I was the one the attorney representing the campground's interest was going to come after to dispute the cause of the fire (by casting doubt on the quality of my investigation). The attorneys for the plaintiffs would question why I issued the citation to the individual instead of to the campground owner since, in their opinion, he was clearly an agent of the campground owner. So I was in the crosshairs, times two. I was quite comfortable with the facts and confident in what I had concluded. But skilled attorneys have a way of asking questions and coming at the same issue from different angles, which can generate confusion.

The serpent's question to Eve in the garden was a twist on what she'd been told. But it was enough to confuse her, with disastrous consequences. So, I had to watch for similar twists and be as cautious and deliberate in responding to their questions as I was in writing my reports. *That shouldn't be a problem*, I thought. I also knew where my weakness was, and that was the fact that I didn't do my formal investigation until several days after the fire, when the weather had significantly altered the indicators.

On the day of the deposition, I was told to plan for one to two hours in the morning, beginning at nine o'clock. Wow! Two hours of questioning? I didn't think there could be two hours' worth of questions to ask on such a straightforward matter. I was to find out otherwise.

I sat at the end of the long table where the deposition took place. My attorney was immediately to my right, and the seven or eight attorneys for the plaintiffs' insurance companies were scattered around the long table. Two chairs to my left was a middle-aged man with fiery red hair, and it became clear that he was going to ask most of the questions. He represented the campground owner's insurance company, and it was his job to blow holes in my testimony, which was key to identifying the origin of the fire, its cause, and broadly implicating the campground.

He looked to be a worthy adversary. Everything was in place. His suit jacket was perfectly tailored. The pleats in his pants were sharp. The shine in his shoes was brilliant. His tie matched perfectly and was tight to his neck. His mannerisms reflected the same supreme confidence. I would have to be very careful with him. He was kind enough, but he had a sharp look in his eye, and I knew right away that however he approached me, today he was not my friend.

Two hours, huh? It was after ten o'clock, and the questioning had not even gone beyond my background. They asked about my education. They asked about my DNR training in Tomahawk when I was hired. They asked about my twenty-plus years of Department of Justice law enforcement training. They asked about my fire experience in detail. They asked about my background in investigations, both fire and otherwise. They asked me to give some examples of the sorts of fire events I had investigated. I knew that was an opportunity to trip myself up, and I tried to provide sparse details. But they rooted out an extraordinary amount of detail on that case (that *I* brought up). We had used up most of the two hours, and no one had even mentioned the fire at Crystal Lake. This was going to be a long day.

I was asked if investigating such fires was part of my responsibility as chief ranger. When I responded that, normally, such fires were investigated by local rangers, I was then asked if I was acting outside of the scope of my authority. Of course I wasn't, and they knew it. But they wanted to give me every opportunity to say something stupid. I tried hard to disappoint them. It was that sort of day.

At one point, I mentioned some burn indicators I observed a couple of hundred yards downwind from the origin as being part of the reason I believe the fire burned in that direction. I described the scorch marks as being higher on the north side of the trees than on the south side. They smelled blood.

"Well, that would indicate that the fire there came from the north and not the south, wouldn't it?" my redheaded rival inquired.

"No, just the opposite."

With faux humility, he went on. "Well I don't know much about fire, but it seems perfectly logical to me that the side facing the fire would have the most burn on it, wouldn't it?"

Okay. You want to talk burn indicators? Fine. We'll talk burn indicators. I know this stuff, and you don't. My attorney was concerned, but I powered forward, now freshly energized. "No, that's not how it works. When a wind-driven fire hits an object, like a tree, there is an eddying effect that takes place, and the flame wraps around and climbs the back of the tree." I used my hands to demonstrate what I shared in my lecture. "Therefore, the scorch mark on the back of the tree, away from the direction where the fire came from, is higher. That is consistent with the stump pile as the source of the fire."

My answer surprised some of them. That was the end of that line of questioning and was perhaps the first time they became concerned that I just might know what I was talking about.

Fifteen or twenty minutes later, I was asked to try to specify the time when a particular individual had told me when they had seen the fire burning north into the woods.

"When did you say he told you about seeing the fire burning in the woods? Tell me again; what time was that when the woods burned?"

Ah, the double-question confusion. Got it. "Are you asking me what time the fire burned, or are you asking me when it was that I spoke with him about the fire burning?"

"Well, let's do both, okay?"

"Okay. I know what time the fire burned because of the various dispatch records and through discussion with our folks who initially responded to the fire."

"Alright. What specific time did you say it was when he told you he saw the fire burning to the north?"

"I never said he told me he saw the fire burning to the north."

"But you have told us here that you are sure the fire burned in that direction, didn't you?"

"I did. But—"

"I thought you said that you spoke with this man about the fire."

"I did. And that conversation took place early in the afternoon, three days after the fire. But I was asking him about who he observed doing what jobs around the campground. I never asked him about seeing the fire, and I don't believe I ever said that I did." We were eye to eye.

I had to listen very carefully to the exact wording of their questions, and I had to think through very carefully the exact wording of my responses so as not to either contradict anything I had said previously and not to let them lead me further down a path that I had never started down in the first place. It was pretty nerve-racking, and there was a lot at stake.

We took a lunch break and started in again. Most of the break was spent with my attorney counseling me on what to watch for, but we were growing more appreciative of one another and were pretty comfortable with how things were going. Most of the afternoon involved posturing and covering the same ground, looking for cracks, inconsistencies. It went well. Eventually, four o'clock rolled around, and the question of time came up.

"Mr. Anderson, would you like to stop for the day and pick up again tomorrow?"

"I have a pretty full day tomorrow. How much more time do you anticipate needing?"

My attorney piped in. "There can't be too much more to cover here."

"Just a few more questions." They were retreating.

"Let's get it over with," I smiled, and there was chuckling all around the table. Their well had run dry, and I think they hoped that overnight they might be able to dredge up something else. But there didn't seem to be too much enthusiasm for that. Within twenty minutes, the questioning was done. It had been a long and intense day.

Eventually, it went to trial. My biggest concern, my greatest point of vulnerability, I thought, continued to be the delay in the formal part of the on-the-ground investigation. I thought that was the greatest opportunity to raise doubt with the jury. That didn't come up materially during the deposition, but I thought perhaps they were "saving it" for the trial. Being a trial, in front of a judge and jury, the questioning was likely to be much less rambling and more to the point.

My attorney was prepared to ask me a few questions to establish me as an expert witness, but before the second one, the plaintiffs stipulated to my credentials, and the "friendly" questions were over. One of the plaintiffs' attorneys asked me a few relatively innocuous questions. Then my redheaded friend took the floor, and I was fairly sure that he was going to pin me down on that delay in the investigation.

He asked a few questions about my experience. Then he asked some questions about the likelihood of such a long holdover in the brush pile like that. I dispelled that with the observation that we'd found hot spots in the pile *after* the fire. From there he went to questions about some limited lateral spread in the fire. It was a point from that initial lateral spread, I attested, that the fire jumped the two-track road in the sand, and the fire took off. He asked me some specific questions about what I saw on the ground that led me to believe that the fire had burned sideways and then jumped the road.

"Mr. Anderson, what specifically did you see on the ground that led you to believe that the fire jumped the road at this point?"

"There were indications on the burned fuels on the south side of the road, upwind, the brush pile side, that they had burned from a slow-moving lateral fire. The fuels, the grass across the road to the north left heavier stubble, associated with a running, wind-driven head fire."

He was using a blown-up sketch I had made of the origin. "So the brush pile was here, and the head fire, you say, started over here. Is that right?"

"That's correct."

"How do you know that the fire jumped across the road here?" he asked, pointing to the sketch.

Where was he going? "Because of the indicators I saw."

"Was there any fire across the road directly across from the brush pile?"

"No, nothing directly across from the brush pile." He was skillfully narrowing down to the question about the delay and the compromise of the evidence by the rain.

"Well then, how can you say that the brush pile is the cause of this fire?"

"As I said, because of the indicators on the burned materials in the laterally burning fire that burned to the east of the brush pile before it jumped the road."

"It was a couple of days after the fire that you conducted this investigation?" *Uh-oh.*

"I looked at the area on the evening of the fire, but I did not take careful photographs or sketch the scene until three days later when I conducted a careful examination of the site."

"So the photographs we've looked at were not taken until three days after the fire."

"That's correct."

He's got me now. Here it comes—the question that's going to force me to acknowledge that the scene had been degraded by the rain when I did my full examination of the scene. He's going to ask me how I can be sure how the fire started if I delayed my investigation until three days later, after rains had essentially washed out the site. I have to be truthful. How could I have avoided this? Could I have squeezed in some better pictures in the fading light on the evening of the fire? Should I have canceled my activities on the day after the fire so I could do the investigation sooner? How much will my disclosure influence the jury against my testimony? And will that impact the award the plaintiffs rightfully deserve? How can I characterize my response to minimize the impact of my delay?

Red approached me, looking at his notes. He glanced up at me and then back to his notes. *Get ready, Blair!* After a long pause, "Your honor, I have no further questions of this witness." And he returned to the plaintiffs' table and sat down.

What? It's over? You're kidding me! In my disbelief, I still saw my opening. I looked at the judge, who nodded, and I wasted no time in exiting the witness stand.

But how did he miss what he had? How did he not see that he had only one more question to cast significant doubt on my conclusions? I pondered that much in the days that followed. All I could conclude was that most people, even sharp attorneys like Red, really understand very little about fire, wildland fire, and how it happens. On that day, that worked to my advantage.

A couple of weeks later, my attorney friend Red called and asked if I would be willing to serve as an expert witness for him in future cases regarding forest fires. I thanked him and observed that in my current job, there could be a significant conflict of interest and that while I appreciated the offer, I would have to pass. I was both humbled and encouraged that he had asked.

Later, in the appeal process, which did not involve me, my attorney made a very insightful argument to the appellate court related to the question of agency and the extent of damages that could be awarded in such a case based on a hundred-year-old statute. The court agreed with her reasoning, and a precedent was set for future damage awards resulting from forest fires. It was a brilliant bit of work on her part.

SEARCH AND SEIZURE

My fascination with constitutional law generated some practical applications in the field. The law that most influenced how I worked at a fire had to do with my authority to access the fire scene. When a fire is burning, it is considered an exigency. As such, an officer, like me, or an emergency responder, like any fire personnel, has the right to go onto private property without specific authorization to address the emergency. Once the fire is out, however, the exigency no longer exists, and an officer needs either landowner permission or a search warrant signed by a judge to be able to go back onto that private property to investigate.

Many wildland fires, especially later in spring, involve woods and heavier fuels, generating the need for protracted mop-up, which can sometimes carry into the following day. It is an appropriate practice to do the best job of mop-up only until it gets dark, after which it becomes significantly more dangerous to be in the woods—tripping over logs, getting branches in the eye, and so forth. Under those circumstances, I would tell Louie the fire was contained but not out (an accurate status), and we would plan to come back the next morning to continue mop-up or to better assess what may still be smoldering. In those circumstances, I am free to return to the property to do additional suppression work and, incidentally, to do whatever additional on-the-ground investigation I need to do.

But in a different situation, where the fire may be in lighter fuels, for example, I quickly learned to do all of that sort of work while I was still on the property from the initial suppression response. If I got called away to another fire in an emergency, I would characterize the fire (to Louie) as contained but not out because I had not had an adequate opportunity to inspect the whole perimeter of the fire. Therefore, I could return legally, and for bona fide reason, to assess the perimeter and gather whatever other information I needed on the ground, as long as I was there.

It is right that law enforcement officers should not have a blanket right to come onto private property for whatever reasons they may have. It is also right that emergency responders, including law enforcement officers, do have the right to come onto private property in response to an exigency. Good law enforcement officers make appropriate use of those opposite notions to do their jobs as respectfully and effectively as possible.

Off-Roading

While I had the authority and the tools (lights and siren) to require vehicles to pull over in fire-related circumstances, I almost never used them. On one occasion, I arrived late to a dead-ended fire scene, and a vehicle was leaving that fit the characteristics of the one being driven by the person suspected of starting the fire. I turned around immediately, threw on my lights, and pulled him over, holding him there until someone knowledgeable from the fire arrived. But a more interesting "pull-over" had nothing to do with fire.

I was in my office at Tower Hill, probably doing paperwork and monitoring some interesting traffic from Iowa County on the scanner. The sheriff's dispatcher was sharing with a deputy information they had received of a vehicle driving erratically, traveling north on Highway 23, which would have been toward Tower Hill. It was generally described as a camper unit on the back of a pickup truck, with additional details given. Less than two minutes later, I turned around in my desk chair and saw through the window the vehicle described drive slowly past the entrance to Tower Hill on County Road C. I immediately left the station in my fire truck to monitor the vehicle's progress, contacting the Iowa County sheriff dispatcher with my information.

The vehicle was going so slowly that I was able to catch up to it very quickly. We were on County Road C, a little used county road that served as a shortcut connecting Highway 23, a north-south road, with Highway 14, an east-west road that connected Spring Green to Madison and other towns to the east. Highway 14 is a heavily traveled major artery a mile and a half from Tower Hill. I observed the camper all over the road. It nearly drifted off the shoulder to the right, where it would undoubtedly have rolled, and another time crossed the centerline so far it forced an oncoming vehicle into the ditch to avoid a head-on collision. I was in contact with Iowa County, providing an ongoing, blow-by-blow account of the driver's actions and encouraging them to get a unit there as quickly as possible. I expressed fear as to what may happen if the vehicle reached Highway 14.

Since he was only going about ten miles per hour, we still had three or four minutes before he would reach Highway 14. But when no police vehicle appeared by the time we reached the stop sign at Highway 14, and with two more close calls with oncoming traffic, I decided public safety dictated my doing whatever I could to get the vehicle off the road before he failed to navigate busy, high-speed Highway 14. So, I turned on my red lights and siren and pulled up immediately behind the vehicle. Fortunately, there was a break in traffic on Highway 14, and the driver slowly pulled the camper over to the shoulder in less than one hundred yards; it was more like the vehicle ran out of momentum than actually stopped. I was clearly outside my scope of authority. If there had been a problem, if the driver decided to bolt and an accident resulted, I would have had trouble. But with what I had witnessed, I thought the risk was justifiable.

An officer never knows what they'll encounter with a vehicle, especially one like this: a pickup truck with a camper on the back with several windows, better for seeing out than seeing in. Not having a sidearm, I was hesitant to approach the vehicle and was uncertain how to proceed now that I had him pulled over. Over my loudspeaker, I told the driver not to move the vehicle, turn off the engine, and stay in the vehicle. As I anxiously pondered what to do next, a squad from the Spring Green Police Department pulled up behind me, red and blue lights flashing.

We spoke briefly, and I watched as he approached the vehicle, cars racing by, and asked the driver to get out. He tried to perform a field sobriety test, but the driver could hardly stand and was arrested, cuffed, and installed in the back of the squad car. Next, and dodgier, the officer went to check the camper. When he opened the door, a middle-aged woman fell out onto the ground, so drunk she could not get up. Soon, an Iowa County squad arrived, and the situation was secure.

The next day, my boss spoke with me about the incident, confirming that I was cognizant of the risk I had assumed. We were on the same page. The sheriff himself also called to express his appreciation for the assistance. In this case, it all worked out well, but I knew there was a potential for disaster anytime I took a controversial initiative like this. Clear, soundly based decision-making is crucial under such high-stakes circumstances.

Chasing a Bug

I don't honestly recall how the idea first crossed my mind that I was dealing with a serial arsonist. There had been one or two fires in a particular township in Dane County, just off the road, whose cause was not immediately apparent. Those happen periodically, whether from a car's degrading catalytic converter, hot exhaust from a tractor, or other random situations.

Perhaps it was the third fire in the same general forty square mile area that got me thinking about arson. When I shared the concern with the fire chief of the department that provided protection in that area, he reflected that there had also been a couple of fires by structures that were similarly mysterious.

He was from a family that had been in the area for several generations, so I thought I could trust him. From then on, we stayed in close touch and were more vigilant about activities in the township.

Fire season ended soon after that, with things greening up and the rains of May curtailing reflections on arson.

The little series of fires had not crossed my mind again the following spring until there was a page for the same fire department in the same general area early one evening. I responded from home, even though it was a minimal fire. The fire department was heading back even as I left the ranger station.

I met the chief at the station, and we went for a drive, noting the locations of all the previous fires that had stimulated our thinking. The chief had offered to drive in his personal car, but I suggested we take my lime yellow truck to project a higher profile and possibly discourage our possible arsonist friend from further activity. We took note of the locations' characteristics, curves in the road, proximity of homes, elevation, points of view, topography, and so forth. We noticed some consistencies, some patterns, and speculated about the particulars driving someone to select those specific locations to burn. Generally, they were out of direct view of any home. There were curves in the town road in both directions from the site of the fire. They were in a place where, once the fires grew in size, they would be viewable from another road. If it was, in fact, an arsonist, and that seemed increasingly likely, it surely appeared to be somebody local who knew the roads, traffic patterns, and likely pathways of response of local fire and law enforcement resources.

Things were quiet for a week and a half before the next episode. This fire was in a new location but in the same general area and had most of the same characteristics that we had noted previously. Our examination of the fire origins became much more thorough, and the entire fire department was especially careful to protect likely areas of origin from disruptive vehicle and foot traffic and random water spray to protect any possible evidence. We looked for something as small as a matchbook, or even a match, or evidence of a cigarette-based fuse timer. Such a device would allow the arsonist to leave the scene before the cigarette burns down and ignites the attached match, initiating the fire. A cigarette leaves an identifiable ash pattern when it burns in place, though it is delicate and easy to destroy. We found nothing.

We concluded that there was wisdom in carefully enlarging the circle that knew about our concerns. The chief knew who the knowledgeable, discreet locals were into whom we could put our confidence and the short list of those who bore our scrutiny. We agreed it would be helpful to draw the Dane County Sheriff into our efforts. I suggested township officials might also be useful. He paused. "I know the town chairman pretty well. We can tell him. But, for now, let's wait on anyone else in the township."

I was intrigued. "Do you have someone in mind?" I asked.

"I'm not sure," he replied. "Let me think about it for a couple of days." It sounded like he wanted time to think through not so much the "who," about which I believed he was more certain, but rather the politics and machinations of the "how," if he *was* right.

I trusted him and his judgment, so I let it be. A few days later, the chief called me and said that one of the deputies who frequently worked on that side of the county was interested in hearing more about our concerns. So we set up a time to connect.

The deputy played it pretty close, not sharing any specific suspicions, but it was clear that he had some ideas of his own. Again, I didn't press. This was all pretty new, and the allegations we were contemplating were pretty serious. The arsonist had not, to this point, caused serious damage or injury, but that is always possible with any emergency fire response, regardless of the seriousness of the underlying incident. I knew that another fire or two of the same ilk would loosen tongues significantly.

On cue, our guy touched off a patch of grass at the side of a town road one cool evening. He had to know that the circumstances that night would not result in a significant fire. But did that mean that in his heart he really didn't want to do any damage, or was it more that he was just toying with us? Or both? Again, we examined the origin without finding anything noteworthy. This time, though, the sheriff also responded with a squad. Several of us spoke with the people who lived on the road as far as the next intersection in both directions. We asked them if they had seen anything but did not share our suspicions. While this did not yield any helpful information, it raised the discussion and concern level at the local eating and drinking establishments. For someone wanting to start fires and not get caught, the stakes had been raised.

There was one additional fire that spring that may or may not have been the work of the same person. It was close to the same area. Again, there were no indicators of the cause, but one or two of the other characteristics of the previous fires were a little different here. We never resolved the cause of that fire. But after our big presence in the neighborhood after the last fire, there was no additional activity that spring.

The following spring, however, started substantively. Perhaps the intervening ten or so months provided the opportunity for our bug to reflect on the challenge and plan how to initiate these fires more carefully. Two fires, the trademark of our violator, happened within the same week in late March. But again, we turned up no evidence. Still, word that someone was setting fires started to get around, and most people weren't aware of the fact that the timing of the fires was such that they were unlikely to cause significant damage. So, a certain amount of ambient fear had developed. People were watching.

We caught a break, of sorts, a week or so later. A piece of information worked itself through the local grapevine: a resident in the area had come upon a car stopped by the side of the road and an individual out in the ditch. As soon as the resident had passed, the individual quickly got back in the car, turned around on the narrow town road, which was not an easy task, and took off quickly in the opposite direction. It was dusk, and they were not able to provide much information characterizing the vehicle. But smart money told us that this was our guy, and we figured that close call had rattled him.

I asked the chief for another get-together, and I prodded him about his previous hesitation to share our information with others in township government. After a brief, reflective silence, "Okay. There's a patrolman who works for the township," and he paused.

"Do you suspect him?" I asked.

"No," he responded. "He's not that sort of guy, I don't think. And he's older. But here's the thing," he continued. "He has a son, who always seemed," again he paused, searching for the most proper characterization, "a little different." I knew he was working hard to find an appropriate way to share some potentially damaging information, so I waited quietly. After another long pause, "I don't know," he thought out loud. "It's just that he doesn't seem to have a lot of direction. Seems like he only works sporadically, and I think he still lives at home," he concluded.

"Interesting. How old a fellow is he?"

"Well, let's see, I think he was in my little sister's graduating class, so he must be mid-thirties or so."

"And he's still at home, huh?" I offered.

"At least sometimes he is."

"So, tell me about his dad."

"He's been the patrolman here for a long time, a quiet sort of fellow, and they keep pretty much to themselves. I really can't see him doing anything like this."

"What all does this patrolman do?" To me, the term "patrolman" had always carried a law enforcement connotation, so I asked about that. "Enforcement work?"

"No, nothing like that. The usual stuff. Mowing in the summer, plowing in winter, sometimes some grading, town roads, minor repairs, the usual sort of stuff. Just trying to keep things working right in the township."

We were going to visit him but decided it would be best to share our thoughts with the sheriff and have the chief make discrete contact with the town chairman first. I was unable to make the meeting with the Dane County deputy, but I was told that he was less than shocked that this individual was a subject of our focus.

The law enforcement community in Dane County and in the city of Madison had experienced some limited interactions with him. Nothing serious, but disagreeable little noncompliance issues, like driving an unregistered car and failing to pay parking violations. Those issues did not point to arson, but they did portray an individual at least somewhat comfortable with ignoring what he apparently thought were minor issues of local regulation. With our confidence growing, we arranged a meeting with the town patrolman, the father of our suspect, the next day. But that evening, our firebug struck again. This time the fire was set a couple of hours earlier. It was warmer and dryer, and this fire made a little bit of a run toward a barn. The fire department's response stopped the fire before it reached the barn, but had they been an additional three or four minutes later, the barn may well have become involved. So it was close. The characteristics of the location of the start pointed toward our arsonist.

Beyond that, we did not know what to think. The fire was not started with the intent of avoiding damage. To the contrary, it had the potential to do much more damage, and it felt like he, too, had heard the rumors and was taking up our challenge. Had he heard that we were going to have a conversation with his father if it was, in fact, him? If so, did he view that as an exciting challenge or angrily as a direct threat? Either way, the game had changed.

Arson can be the solution for a variety of needs. For some, it's a business, a paid service the procurer uses to take advantage of insurance coverage. With a skilled firebug, investigators know the fire was set, but nothing can be proven. So, the insurance company pays off on the coverage, and all their policyholders lose. A revenge motivation occasionally leads to arson, and arson is sometimes used to cover up another crime, like burglary or fraud.

After that, reasons for starting a fire get a little hinky. Some find it satisfies a desire to exert control over others, such as making a group of volunteer fire personnel drop whatever they are doing to respond to their fire, which may have been a component in what we were looking at. It can be an overdeveloped sense of playfulness. At the far end are the cases found in the *Abnormal Psychology* textbooks; let your imagination wander.

Before that evening's response had ended, two Dane County squads had joined the fray. It turned out they had done some investigation of their own into this individual's activities and had been making a limited effort to monitor his whereabouts. He had apparently changed his usual pattern for that evening, and they lost track of him. Again, our confidence swelled, and any draw we had felt toward discretion in our investigation had evaporated.

We decided to delay our meeting with the town patrolman so we could better organize our information, considering the new insights on the son, and make the most of our time with him.

The chief and I met again the next day. Keep in mind that his role as the chief is volunteer and that he has a full-time job. Employers, especially rural employers, understand how much that willingness to serve means to a community, and they tend to be generous with freedoms given to employees who help in such ways. Still, it is an additional source of stress for someone who chooses to volunteer in that way.

I reorganized our information and shared it with him the next afternoon. We decided to add a couple of questions to the previous list and arranged a get-together with the patrolman the following afternoon near the end of his workday. The chief shared the plan with the town chairman, who was acutely interested in the problem and whom he had kept loosely abreast of developments.

We arrived early and found the patrolman ready to close the shop. We did not know if his plan had been to sneak out early to avoid the awkward discussion. In his office, we began to lay out our inquiries to him. With each additional question, he began to realize the extent of what we already knew. While he certainly didn't incriminate his son, neither did he take advantage of any opportunity to robustly defend him. We came away without any evidentiary epiphanies, but we did have an understated confirmation that we were looking in the right direction.

A week and a half later, there was another fire. Small and inconsequential, and set to be so, it seemed like a pitiful whimper. Still, we responded aggressively, going to the patrolman's home that evening and asking him about the whereabouts of his son, our suspect. He said he didn't know, and we believed him. But in the process, we met his wife, the mother.

There's a Filipino saying: "If you want a man to do something, ask his mother." It worked with Jesus at the wedding. And a corollary to the idea seemed to have some benefit here, too.

Mothers and fathers can make good teams, filling different roles in the dynamics of a family. In this case, our visit that evening may have been Mom's first wind of our suspicions. And that may have predicated the most powerful influence on the situation yet.

We never again had another fire that had any indications of having been started by our arsonist. Perhaps "Mom's" expressed pain at the possibility of her son getting caught in such activity was the motivation to stop. We didn't know. Although we were still a long way from having a prosecutable case, our actions seemed to adequately serve our objective of stopping the problem.

The overall result left us, as well as the sheriff, who promised to continue to keep a watchful eye on him, a little dissatisfied. We were never able to make any public announcements. But it brought the problem to an end, and the concern that had filled the community slowly dissipated. I, we, had done our jobs well, at least well enough to bring this threat to an end.

Paranoia?

One spring day in 1995, we were just a couple of minutes from closing the stations at five thirty. We stayed open an extra hour, even though there had been rain a couple of days before, and the humidity was expected to stay high, so the fire danger was not as great as it frequently was at this point in spring. The plans to close had been broadcast to all the ranger stations and fire personnel over the radio on our usual frequency, one that a lot of folks in scanner-land monitor.

It had been an unusual week, as the bombing of the federal building in Oklahoma City had only occurred a few days before. With surprising ease and speed, the FBI was able to identify and take into custody the suspect in the bombing, who was subsequently found guilty and executed. It was a surprise to the public to learn that he was not a foreigner but a lifelong American and a veteran of the U.S. Army. It also came out that he had some association with a militia group that had voiced concerns about the extent and nature of government activity in late twentieth-century America.

When the phone rang at the ranger station at 5:29, nothing seemed unusual. I answered it, and the caller told me that a neighbor was burning on Highway N north of Plain and that he was concerned that the individual's brush pile might get away and threaten some of the caller's buildings. He also said that his neighbor routinely burned illegally, and he was hopeful that contact with him might make him stop. When I asked for a name, the individual said he wouldn't provide it but gave me a general location and hung up. Even this somewhat mysterious behavior wasn't unusual because neighbors often didn't want their neighbors to know that they had reported them.

I called Louie on the phone and told him about the call and that I thought I should probably go up and investigate. He agreed to stay on the air while I did that; he'd monitor my radio traffic and stop and dispatch for me from the Spring Green office on his way home if I needed it. So, I went to straighten out this illegal burner.

It took me twenty minutes or so to get to the road I had been told about, and by that time, Louie had arrived at Spring Green and contacted me on the radio there. I informed him of my location as I turned onto the four-mile-long road the caller had told me about. The burner was supposed to be about three-quarters of a mile or a mile up the road, on the right, or north. The roads there twist and turn and go up and down through the hills, so the smoke column I sought would not jump out at me like it might on level ground. As I reached the three-quarter-mile point in the road, where I expected to find fire, I slowed, searching more intently, but I still did not see anything. Dry brush piles burn ferociously when they are initially lit, and there is much fire and smoke. But they burn down quite quickly, leaving a hot bed of coals that combusts more completely and doesn't generate much smoke. So I watched carefully as I proceeded down the road. I reached a mile, then a mile and a quarter, then a mile and a half, and still nothing. I went a little farther and turned around, driving back slowly past the area in question, thinking the different perspective might help spy the pile. Nothing. I turned around and went back a third time, continuing the remaining couple of miles to the next intersection. Still nothing. I turned to head home.

I reported to Louie what I had found, or what I hadn't found, and released him to go home. It was on the drive home that the bogeyman began to creep into my thoughts. I had become pretty well convinced that there was not a fire along the road, anywhere. My thoughts started in the direction of why someone would call and report such an incident if there was none. As my thinking drifted toward Oklahoma City and the resistance/opposition toward government employees, it occurred to me that at that late hour, just before we were planning to close, that remote location would be a good place to ambush a government employee. My own colleagues, apart from Louie, had all gone home. It was right around the time of the shift change at the sheriff's office in Baraboo, thirty miles away, so the deputies on duty would all be up there, and it would be a long time before any other law enforcement officer could respond to my location. As these thoughts, irrational or otherwise, washed over my thinking, I began to get the sense that I didn't know what was going on around me. Paranoia has an extraordinary capacity to advocate for itself, and it was in full swing. I was suddenly very glad to be heading home. When I got to the ranger station, I called in that I was out of service, as was our custom, to advise Louie and the supervisors that all was well. But before I went home, I wrote a detailed incident report to document what had happened in case something similar happened again.

Almost thirty years later, I still have a vivid memory of how that felt. As I reflected on the innocuous events of that evening over the next days and weeks, I developed a new appreciation for the exposure that law enforcement officers have when they delve into unknown circumstances based on unconfirmed complaints related to people with unknown motives. It takes a particular constitution to deal with that day in and day out and not become neurotic. That's difficult for people to understand without even the little exposure to law enforcement that my job afforded.

Suspect

For some reason, I was home alone on a cool November night. We were dog-sitting a friend's pooch, and I took him out for a walk. On an otherwise sleepy evening, my attention was drawn to a glow in the sky to the north. It looked like it was close by, and I could not think of any explanation for it that wasn't ominous.

I hustled Manchester back into the house, got in the car, and drove around the corner to the neighbor's, which was a mile to the north. As I crested the rise and the farm came into view, I could see smoke and occasional flames coming out of the top of the barn on both ends. I ran to the house and banged on the door. When there was no response, I tried the knob, and it was unlocked. I went in and used the phone to call the sheriff to page out the fire department.

Going to the barn to see if anyone was there, I was surprised at how quiet and calm it was on the lower level of the barn. There was no indication that there was a fire above. But the bad news was that all the cows were in their stanchions. I walked to the center of the barn and tried to solve the release mechanism to free the cattle and get them out, apparently unaware of the fire above. This is where being a city boy was a distinct detriment. I'm sure the mechanism was easy to work for someone who knew what they were doing, but despite my desperate efforts, I could not open the stanchions. After a couple minutes of trying, some embers began to drop through from the floor above. Despite the consequences, this wasn't worth me dying over. I got out of the barn.

It was an exceptionally helpless feeling to see the fire spread to the lower part of the barn and be unable to help the terrified animals.

The fire department arrived and began working on the barn. But it was clear that with the extent of the fire's development and a heavy load of dry hay in the barn, it was already lost. There was nothing to do but focus on protecting structures that would be exposed to the growing heat from the barn fire. I could do little but watch.

The most dramatic moment occurred twenty minutes later when my neighbor came home and frantically tried to get into the barn, then fully involved, to release the cattle. A couple of us almost had to tackle him to keep him from going in. As hard as it was to see the cattle in their panic, it was worse to see this man watch his barn, his cattle, his livelihood, his way of life, go up in flames.

The rest of the night was spent hauling water from town to a holding tank the fire department set up in the yard from which they supplied the hose layout used to contain the fire. I was able to help by hauling water using our tanker through most of the night. It was a long, sad evening. In the morning light, ribs could be seen protruding from the burned carcasses of some of the cattle.

A couple of years later, another barn burned, this one over the hill on the other end of the road that my neighbor and I lived on. Similarly, it was a loss, and another night was spent hauling water.

It's not unusual in such a total loss that a definitive cause is not identified. It's often assumed that old wiring in an old barn failed, or a bearing in an old pump overheated, or some other equipment-related issue was the cause. If I ever knew, I no longer remember what the presumed cause of either fire was. But I would be disappointed in the county detectives, some of whom I knew fairly well, if my name never came up as a potential suspect. The fires were both close to my home, they were of a similar type, in barns, and there was no clear cause. As an investigator, I know, like everyone else, that occasionally there are people in the fire business who enjoy their work enough to create a little more of it.

Apparently, consideration of me as the source of the fires never rose to the level that I was contacted. I'd like to think that it was dismissed immediately. But those thoughts never go away completely. And it was strange to know that such an unspoken notion, however slight, was out there. Sometimes, circumstances make those suspicions unavoidable, but it is very difficult to be even momentarily considered capable of something evil by people you know well and work with, whether it's true or not.

FORESTRY

REFLECTIONS IN FORESTRY

My job leaned heavily toward the fire program. My sub-area was one of the three largest (of fifty-eight) in the state, and we had more fire departments and more fire wardens than almost any other ranger station. Half the years I was there, Spring Green had more fires than any other station as well, mostly because we have such a large population, and human activity is the cause of the vast majority of wildfires in Wisconsin. So, 70 to 80 percent of my job involved fire.

As a result, my forestry assignment was generally of an assistive nature, usually targeted at supporting county foresters in one of the four counties where I had fire responsibility. One year, however, it was decided a big chunk of my forestry work for the year would be to serve as the forester on the master plan committee for Mirror Lake State Park in northern Sauk County. Master plans were developed or refreshed every fifteen to twenty years for state properties to ensure that they were serving a broad range of areas: wildlife management, water quality, recreational use, endangered resource protection, aesthetics, and forestry. Master plan teams were comprised of people who had expertise in those specific areas, plus property management personnel from the discipline responsible for the property, in this case, Parks. (Besides Parks, Forestry and Wildlife Management had primary responsibility for other properties around the state.)

As the forester on the team, my job would be to look at the vegetative cover over the entire park, identify and map various stand types in the park, and develop forest management recommendations for each of those stands over the next twenty to fifty years, which would maximize benefit, but in the context of the primary use of the property. So, for example, even if the best practice for a particular stand might otherwise be to clearcut it, a modified version of that is prescribed because of the nature of the use there. Mirror Lake State Park was about 2,200 acres, so there was a long process of simply gathering the data about the vegetation on the ground. After processing that data, I'd identify distinct stands and map where those stands were. I would then decide the best way to manage those stands and write up a detailed plan that included specific activities to be done there and when those activities should take place.

It was my kind of assignment. The only deadline was for me to turn in my final report sometime in the fall. Apart from that, I could do the fieldwork and the write-up anytime I wanted. If there was a nice day, and I didn't have anything else specific scheduled, I went to the woods at Mirror Lake. If I didn't want to deal with anything on a particular day and I wanted to get away, I went to the woods at Mirror Lake. If it was one of those glorious early spring days before fire season began, and I wanted to get out, I went to the woods at Mirror Lake.

It is an interesting place with lots of water and very interesting contours. For the most part, the forest types were ones I was thoroughly familiar with. But what was most interesting were some deep, rock-walled ravines that could only be entered from one end. I discovered the first of those on a fairly warm summer afternoon. As I descended into the ravine, the temperature dropped until it was at least fifteen degrees cooler at the bottom. Cool air is heavier than warm air, and the ravine is shaded most of the day and sheltered from the winds that tend to stir air to a uniform temperature in an area.

Even more interesting was what I found when I got down into the ravine—at least two species of trees that I had never seen in the wild outside of northern Wisconsin, at least 150 miles to the north, yellow birch and eastern hemlock. It quickly became clear to me that this was a remnant, of sorts, of a timber type that had prevailed there millennia ago but which could not compete with other species as temperatures warmed, perhaps as the glaciers receded. But down in this ravine, where it stayed cool and protected, these trees and their associated undergrowth, also of a northern Wisconsin type, persevered. It was an interesting in-the-field macro lesson. Though there was little to do with it from a forest management standpoint, I prominently mentioned it in my report as something to point out to park visitors as an interesting example of how microclimate influences how forest-type changes take place or, in this case, don't.

Branching Out

When I bring a case to court, it's because I believe somebody's done something they shouldn't have, and under circumstances where there needs to be significant motivation to not do it again. I am already convinced of their guilt, or I wouldn't take such an action.

However, the court, our legal system, approaches a case differently, with a presumption of innocence that needs to be overcome. Sometimes, that convincing effort is put on a jury. However, there is always a judge involved, and any way their opinion can be appropriately influenced is, in my perspective as an officer, a step toward justice.

Occasionally, a scientifically-based item will pique a judge's interest.

A complaint had been brought to a neighboring ranger station that timber had been cut illegally. The landowner's neighbor had conducted a timber sale, and the cutters came across the property line, through the fence, to cut some additional trees without permission. My neighboring station was short-staffed, and with my experience and interest in law enforcement, which is not common among the foresters, the investigation fell to me, to my delight.

When I initially inspected the property, I found twenty or twenty-five trees that appeared to be across the fence line, thought to have been taken without consent. Fortunately, the logs from those trees appeared to still be on the landing on the neighbor's property and hadn't been hauled to the mill yet. I seized them (placed a seizure tag, supported by the court), a legal action that meant they couldn't be moved. To do so would be, in effect, violating a judge's order—not a good idea. With the logs on hold, I bought myself some time to figure out if there was any useful evidence I could obtain from them.

The particular loggers involved had a well-earned reputation for pilfering trees that were not to be cut or taking trees and not paying for all of them. It seemed like there was a lot at stake for both this landowner and others in the future to get a conviction.

In a case like this, at this stage of the harvest process, the challenge is always to demonstrate that a particular log in one location came from a particular place, or more specifically, a particular stump. People, in general, tend to think that a log is a log, and a stump is a stump, maybe bigger or smaller, or of a particular species, but in no other way distinguishable, one from another. This way of thinking includes most district attorneys and judges, too. But if one knows what to look for, there are all sorts of distinctions.

As I examined the stumps for peculiarities, I found two where a branch, which had begun to grow when the tree was very small, had persisted for many years and was surrounded by many new rings of wood grown in subsequent years. So, the "stump" of this branch appeared in the cross-section of the tree stump. It is unusual for a small branch like this at the base of a tree to persevere for so long, so the appearance of it so far out in the stump was unusual.

This was just the sort of indicator I was hoping for. I went to the log pile, looking for a corresponding branch at the large end of one of the logs. When I found it and examined it, I became positively convinced that that particular log had come off its corresponding stump. There were some irregularities in the "roundness" of the stump and the log that also matched perfectly. Next came the hard part.

I photographed the log and the stump, wrote a narrative of my investigation, and went to see the district attorney. After considerable explanation on my part and some uncertain acknowledgment on his, he agreed to proceed with the prosecution of the case after I promised to go over my explanation as often as he needed it. In the course of our discussion, we developed a mutual "let's get 'em" attitude.

To prepare for the courtroom trial, I had to cut the end off of the log and the top off of the stump with a chainsaw so that I had them in my possession. Pictures would suffice for the trial, but I needed to testify that I possessed the underlying evidence in physical form.

When we got to that part of the trial, I was called as a witness for the prosecution, and the district attorney asked that the court acknowledge me as an expert witness. When someone is an expert witness in a particular area, it essentially means that whatever they say about that subject is not opinion, but fact.

"Have you examined the logs at the landing and the stumps on the adjacent land?"

"I have," I responded, succinctly, noting the yes-or-no nature of the question. I let the district attorney dictate the pace of the disclosure.

"Have you come to any conclusion about whether the logs that were seized have been removed from the neighbor's land, where there was no permission?"

Again, yes or no. "I have."

"Would you please tell the court your conclusion?"

Now the door was open. "I am certain that they were taken from the neighbor's land." That raised a few eyebrows at the defense table and in the jury box, but also in the judge's chair.

The DA was prepared for this and asked that my photographs be entered as evidence. After a few moments for the defense and judge to see them and for them to be marked, they were handed to the jury. He then asked me why I was so confident. I referenced my photographs and explained the implications of the matched branch marks and the noncircular nature of the tree. When I did that, the judge, who had looked at the pictures before, asked to see them again. That was a little unusual. Even more unusual, he interrupted the district attorney's questioning and asked me a couple of questions directly about how trees grow. "Very interesting," he commented. The jury took note of his response.

That was a tough blow for the defense, and they were powerless to offset the judge's obvious interest in the incriminating evidence. The credibility I gained from the judge's response meant that the defense attorney could not question me roughly, skeptically, without incurring the wrath of the jury and possibly even the judge.

Before the end of the day, the defendants pled out to a lesser charge and agreed to pay a fine. More importantly, they had to pay the offended landowner twice the value of his timber, as provided for in the law, and which I got to establish as the expert witness.

It was a good case to win, and it was an interesting way to win it, landing on some basic biology that kindled a judge's interest.

Poor Harold

I had known Harold Kruse since I started working for the DNR. He and his wife, Carla, had long served as emergency fire wardens (EFWs), a voluntary role as a local source for burning permits. In the old days, those positions had also served as fallback firefighting labor. More recently, EFWs tended to be older and were simply available to issue burning permits. I had about fifty

EFW locations in my area, and the Kruses were one of them.

While they had sufficient financial resources, they lived on a property that had been in the family for a long time. Though there was a bona fide house on the property, they lived in a fairly makeshift housing arrangement that appeared to have been originally designed for another unknown purpose. I was only inside once. It was one of those places that appeared to be chaotic and disorganized, with papers piled on most horizontal surfaces, stuff over the backs of chairs, various objects leaning against every corner, lots of hooks on the walls, all used. But I had the sense that they might be able to find almost anything they were looking for. Cramped and minimalist, the home reflected their perspectives on living. They were in their sixties when I first arrived in Spring Green and vibrantly involved in outdoor nature work. They had a large garden and kept some smaller animals, such as chickens, but prairies were their passion. They supported The Nature Conservancy, and Harold served on their property management teams for many of the properties they owned in Sauk County, most of them prairies.

Harold was a diminutive, soft-spoken fellow who seemed always to be wearing the same working style clothes, mostly khaki, and a lightweight Stormy Kromer hat, summer and winter. His way was plain-spoken and honest, and he was faithful in friendship. Beneath his longish gray hair, he had one working eye. The other was uniformly grayed out, the sort of thing that was hard not to look at, even though you know you shouldn't. He was a gentle and humble man, serious and rather humorless.

So years later, when he called me, distraught about some missing timber, I felt an urgency to respond. We were not close, personally; we were different sorts of people. But he was thoughtful and committed, and as a volunteer EFW, he had been helpful to me in my work. The timber was on eighty or one hundred acres the Kruses owned not far from their home, so I agreed to meet him there the next day.

It was winter, and there was at least a foot and a half of snow in the woods, so the going was a little slow, especially for Harold, now in his late seventies. He was so upset initially that I had a difficult time understanding the specifics of what had happened.

Eventually, I learned that he had contracted with a forestry consultant, Jim Carlson, a close, personal friend of mine, whose firm had marked the timber and set up the sale. In the week or so since the site was last inspected, Harold claimed that a large number of unmarked trees had been cut and removed. There was no logging equipment on site or indication of ongoing logging at the time, so I made a very cursory inspection and determined to contact Jim to get more details. Before we parted, Harold told me that the trees that were missing were in a part of the woods that he and Carla had designated as a memorial forest for their late son, an area in which no trees had been marked.

That evening, I called Jim, and he was surprised to hear about the missing trees. He, too, was very concerned and wanted to meet out there first thing the next morning. Before we hung up, he shared some information about who logged the job. It was a name I knew, and not in a favorable way. Jim indicated he would contact Harold directly to ascertain if he wanted to join us in the morning and that he would contact the logger to get a response to our concerns.

The next morning, we met at the site. Poor Harold was only a little less distraught than he had been the day before. No matter how things were resolved, he lamented, the memorial site for his son had been desecrated, and there was nothing that was going to put the trees back on the stumps. While I hurt for him, my focus was on suitable consequences for the logger and for Harold to at least get financially compensated, and then some, for the timber that was missing. Jim was both angry about the situation and apologetic to Harold that it had happened. The snow was too deep for Harold, so he returned home, and Jim and I spent the next couple of hours doing a rough assessment of what was missing.

As we worked through the stand, we found that all the cut trees had been marked with paint, the typical procedure used with timber sale marking. Jim's process involved a spot at the very base of the tree, below where the felling cut would be made, and two spots up high on the tree, one downhill and one on the opposite side, so a standing tree designated for harvest could be identified from any perspective. The upper marks would have left with the logs, but the lower marks would remain on the stumps.

Back in those days, loggers were rarely accused of being uniquely gifted with wisdom when it came to their work. This incident confirmed that notion. (The logging industry has since grown generally more professional.) Jim, having overseen the marking of the sale himself, had some sense of which missing trees should not have been missing. Beyond that, it was not difficult to recognize that some of the trees had been marked after the snow had accumulated around the base of the trees, with splashed paint quite visible in the snow beneath and the basal paint marks well above the base of the trees. Those that Jim's folks had applied the previous fall had to be dug out from beneath the snow.

What had happened was readily apparent. The logger had harvested the legitimately marked trees and then decided to help himself to some extras in an adjacent location, marking them with his own paint in an effort to insulate himself from accusations of stealing. It was a poor plan, well executed, and oh-so-easy to see through.

We learned which mill the logs had been sold to and went to look at them. They were still piled separately, and as we anticipated, most of the butt logs in the pile had the high marks that Jim had painted on them initially.[13] But some didn't. We concluded that it would be easier, logistically, and better, forensically, to estimate the volume of logs stolen from observations on the ground in the woods than to try to get the sawmill to take apart the large pile so we could scale it. Jim contacted the logger to tell him he could no longer access Harold's land (though the damage was already done), and I made a call to my connection in the Sauk County District Attorney's office.

The following week, I went to meet with the assistant district attorney and gave him the flyover of the situation. He was a little

---

13 A butt log is a log from the lowest part of the harvested tree. Since trees taper with height, logs have a slightly smaller diameter on the smaller, upper end than they do on the lower end. A butt log can be identified by a particular flare at the lower end, where the visible part of the tree emerges from the supportive and life-giving root system. That lower log contains, on average, almost 50 percent of the volume harvested from a tree and generates up to three-quarters of the value from the tree, because it tends to be as good or better quality than the higher logs, which are smaller and more likely to be compromised by branches.

uneasy with the situation, understandably, since he didn't really understand forestry and logging.

Most people have little or no experience with selling timber. Even woodland owners may only manage one or two sales in a lifetime. Frequently, a logger or timber buyer knocks on the door with an offer. The landowner, having not thought of the woods as a source of cash, is pleasantly surprised and relies on the buyer to do whatever he thinks is appropriate with the woods. Loggers, generally, are not foresters and tend to look at trees in terms of what they can sell. Foresters understand how trees grow differently and that a smaller tree, because of its slower rate of growth, might be older, mature, and ready for harvest, while a larger tree might have better genetics and still be in its primary growth phase and should be left to grow. The logger has a vested financial interest in getting the most value off of the property immediately. As a result, not only do more trees, and oft-times the wrong trees get harvested, but since the uninformed landowner has little knowledge of the value of timber, once the trees are cut and removed, they accept from the logger whatever they are told the timber is worth. Some loggers are more honest than others. But I know of situations where the unscrupulous ones have paid as little as 10 percent of the market value of the stumpage, and the landowners, in their ignorance, have been delighted.

This was not the first time I had encountered this sort of hesitancy in a district attorney. I explained to him some of the processes and protocols we would use to assess which trees were stolen and make a good estimate of how much timber was missing. Still anxious, he cautiously agreed that I should proceed with that assessment. With the logger barred from the property, some of the urgency abated. It was a couple of weeks before Jim and I worked out a time to go out there and do an exhaustive assessment of the site.

Estimating the value of timber that *isn't* there is a little challenging. Numerous clues provide insight into what the missing trees may have looked like. One, of course, is the quality and condition of the remaining timber. Whatever their characteristics are, it is reasonable to assume that the missing trees were at least similar, if not better.

When it comes to estimating the specifics of a particular tree that isn't there, it's helpful to consider how the logging process takes place. A target tree is felled, and the merchantable part of the tree is assessed, working up the tree from the base until the branching becomes such that it effectively interrupts the presence of logs. At that point, another cut is made, severing the top. Any lateral branches on the merchantable part of the tree, usually quite small, are also cut. The whole tree is then skidded to a landing, where it is cut into logs.

So, measuring the distance between the stump and the tree's top is a credible way to estimate how much merchantable length was removed. Where many trees have been cut and dragged out, it can be a challenge to determine which top goes with which stump, especially if it moved from where it originally landed in the felling process (for example, to skid a more distant log through the area). I've always enjoyed puzzles, so I find this aspect of the investigation stimulating. Another consideration is to look at how many lateral branches were cut off from the missing logs; those branches are left in the woods and can be linked to a particular tree, especially if cutting is lighter. Generally, log quality is inversely proportional to the number of branches removed.

In this case, the timber, which was mostly oak, was of high quality. Jim and I did exhaustive tallying of all the timber that was cut, including what was supposed to have been cut, and we took lots of pictures. Part of the thinking is to have comprehensive information for a potential trial. But it is also important to compare with the volume scaled and paid for by the mill to the logger. Their volume and our volume should be reasonably close, thereby lending veracity to our assessment. As we tallied, of course, we distinguished between those stumps that were marked low by Jim and those that were marked at the snow line, well above the ground. That was an easy distinction, and photographs demonstrated that.

At some point after that, we were able to compel the logger to visit the site. He was not particularly keen on this because he knew we were going to confront him with what he had done. But that meeting took place, and for the most part, the logger's responses were limited to, "I just cut what was marked," and, "No, I don't have no idea why some of the marks are low and others are high," or, "I don't know why some of the butt logs in the mill yard had high marks on two sides of them and others didn't." In his shoes, I might have responded similarly, though I would hope to never put myself in such a pickle.

Armed with this information, I went back to the district attorney and laid it all out for him, with the recommendation that the logger be charged with felony theft. I explained the value of the missing logs, both in their value as stumpage and their increased value as cut and decked logs at the sawmill, significantly higher because of the processing and hauling that had been invested in them. The timber theft law specifically allows for the use of those higher values when determining restitution to be paid to the landowner.

I carefully and thoroughly assembled a comprehensive case file for the district attorney. As I mentioned, the assistant district attorney I was working with in Sauk County was uncomfortable because of the nature of the case, and I wanted to do everything I could to make it easy for him to proceed with prosecution. He agreed, albeit unenthusiastically. Charges were filed, initial appearances made, and a trial set for three months away. There was little more to be done until then.

A few weeks later, Harold called me, again upset. I was finally able to ascertain that something had changed regarding the court case and that he understood the trial was off. What? I immediately called the district attorney.

"Yes," he said. "I agreed to a plea." Okay, that's not always bad. It saves time, avoids the risk that a jury doesn't understand the case, and lets the logger off.

"What are the terms?" I asked.

"They agreed to plead guilty to misdemeanor theft and to pay $2,000 in restitution."

Oh, no! The value of the timber stolen was estimated to be in the $10,000 to $12,000 range, more than enough to classify the theft as a Class D felony, a conviction that could potentially involve prison time and various other restrictions. I had always thought it good policy to contain my temper when working with district attorneys, since I would most likely have to work with them again in the future. But this was too much.

"You can't be serious!" I exploded. "Do you have any doubts that they did what I said they did?"

"No, but—"

"Then why would you let him off with a misdemeanor charge and an $8,000 profit?"

"Well, they didn't make a profit; they're going to have to pay $2,000."

"Yes," I agreed, "as restitution for at least $10,000 worth of timber they stole and sold. How is that justice?"

But the deal was done. There would be no jail time, no probation, no public service. There was no point arguing about it any further. Still, I had to get one last shot in.

"Are you going to tell Mr. Kruse that he's going to get $2,000 for the $10,000 worth of timber they stole from the memorial he set up for his dead son?"

I always took pride in my case preparation for district attorneys, and I had done a good job here. But I had run into a very fearful district attorney. And this time, the logger got off very lightly.

Guess who got to break the news to Harold.

Sleeping on the Job

Occasionally, the travel distance to a work location for forestry would motivate us to work an extra-long day to avoid having to travel a second time on another day. This happened with a large block of work that had to be done at Blue Mounds State Park one fall. It had to be "cruised," which meant that it had to be walked and statistical vegetative measurements taken at appropriate intervals. The essential objective was to find out what was growing where, how big it was, how fast it was growing, what potential future reproduction there was, the quality and form of the trees there, and so forth. This information would be used to establish a long-term management plan for the park to fold together its forestry objectives, as well as recreation, aesthetics, and other offerings. The solo process was walk, stop, observe/measure, and repeat. It made for a pleasant day.

Four of us met there on a crisp October morning, each having traveled the better part of an hour to divvy up the nine hundred-plus acres of woodlands in the park. We were scattered to our respective areas, working away, before eight o'clock. Blue Mounds is a beautiful place with impressive timber, so it was a pleasure to work there, especially on a cool, sunny pumpkins-and-apples sort of day. Much of the park is fairly remote from roads, so once we hit the woods, we weren't coming out until the end of the day. For me, lunch was a sandwich, chewed on whenever I got hungry, and a couple of bottles of water. I worked through the morning and early afternoon and made good progress, confident I would be able to complete my work by the end of the day.

By two o'clock, I was beginning to feel the fatigue of not having been off my feet for six hours. I had reached a gentle south-facing slope, bedecked in crisp brown oak leaves almost a foot deep. I decided to sit down and take a short break before continuing. Those leaves were as comfortable as any recliner that day, and the relief was profound. I still had half of a sandwich left, so I leaned back against the gentle slope, settling into the fluffy litter, and finished munching it down, enjoying the marshmallow sky as it drifted past. Soon, I was drifting with it, lost in a dream. Occasionally, I was minimally aware of a little breeze, but mostly, there was just the warm sun on my face and a sense of weightlessness floating on one of those clouds. Ah ...

My eyes burst open, and I was jarred awake. I was looking up at the same sky, but for a moment, I wasn't sure why. I wrestled with the familiarity of the things I was seeing, struggling to piece together why they were there. Confident that there was an explanation for all this that my mind would soon come to grips with, I waited as if waiting for a fog to burn off so I could see again. For a few seconds, I was completely at a loss. As though on a dimmer switch, my awareness level gradually came back. *Oh, yeah. Blue Mounds.* I assessed the sky for some clue as to the time, but there was none. But I was back in the real world. My little nap could have been ten minutes, or it could have been two hours. Any concern dissipated as I thought through potential consequences. I remembered the other guys and that we had agreed we would each head home as soon as we were done with our respective acreage. So, I might be home late, but I wouldn't be wasting anyone else's time (and I wouldn't get "caught napping"). With that, I refreshed myself with the serenity of my surroundings. All of this happened within five seconds of my awakening. After a couple of deep, restorative breaths, I got to my feet, a little stiff.

Reoriented, I finished the day's work. I was back to the truck at a reasonable time, even finishing before one of my other colleagues. Once I had recalibrated to the correct time, I realized my nap must have been only a few minutes. But both the pleasure of that nap and the confusion at the end of it have stayed with me.

From a Previous Chapter

Sometimes the most interesting feature about a case is not the details of the case itself but who is involved.

Once again, a timber theft in my adjacent sub-area fell to me because of my experience. It was a fairly standard timber theft story. Timber was to be cut on one landowner's property, and when the logger got to the back corner and saw a number of attractive and high-value trees just across the fence in the neighbor's property, he couldn't contain himself.

There isn't as much risk to the logger as it may seem because many of these woodlots are owned by farmers whose focus is on the level land that can be worked profitably. The woods are usually on hills or in steep valleys or draws that are too difficult to farm. Farmers are busy. They work hard, and those remote woods are rarely visited, perhaps only during deer season and maybe not even then. Missing trees will not be missed soon, if ever.

In this particular situation, not only did the logger want to grab the extra trees, but he also didn't want to bring them to the legitimate landowner's pile to be scaled because they didn't belong to him, and the logger didn't want to pay him for them. So, he decided to make a road out of the woods on the seller's property in another direction that would get him within twenty or thirty yards of the county road. That remaining distance crossed a narrow parcel with a house at the other end of the thin strip, almost a quarter mile away. It was occupied by a renter.

The poached logs were removed there.

When I became involved, the job had been completed, the logs had been removed, and the landowner had been paid. After I looked at the neighboring property from which the trees were stolen, I made some measurements, documenting what was missing between each stump and the remaining tree top. I began to look into who had been involved. The landowner did not know the logger's name; he worked through a broker, a middleman, who found and identified timber, and contracted with a logger to remove the product. This broker was somebody with whom I was acquainted. He had logged previously, was familiar with all the unsavory practices associated with logging, and had shown himself to be comfortable with them. So, one evening, I went with my boss, John, to visit him.

We were brought into his living room. His wife was seated in a far corner, quietly working at a table on a craft project, her back to us. It would not have been difficult to overlook her presence in the room altogether. As we stood, John generally explained to him what was going on, that we were looking into some possible inconsistencies, and asked about the volume of logs and board footage that had been hauled from the seller's landing.

The broker folded his arms. "We treat that information as confidential and don't share that with anyone," he answered smugly in a tone designed to bring the conversation to an end. He had made his initial bet.

Silence hung in the air for a while. I had more experience with these sorts of confrontations than almost anyone in forestry, and when my boss didn't say anything, I decided to take a chance. I smiled, sat down, and asked him to do the same. My boss remained standing, as surprised as the broker, I think, at my approach. Our host had lost control of the situation and was clearly uncomfortable. He sat, awkwardly.

"I understand your desire to keep this sort of information confidential. But here's the deal. There has been timber stolen as a result of logging activity on this property. Right now, it appears the value of the missing timber may elevate the theft charge to a Class E felony level. I know you were not out there cutting and may not be in a position to know what may have happened on the neighbor's property. But since you were the middleman in this, and you wrote the checks, that makes you an accessory. This is going to end up in court, and it would be helpful if I could share there your willingness to be cooperative." I'll see your confidentiality and raise you a felony. All in.

That was all I said. I sat, quietly, watching the wheels turn, waiting for him to respond. After what seemed like an eternity of silence, during which we were looking directly at one another, he asked his wife, who had obviously been listening intently to the entire exchange, to bring him "the file." He didn't specify a particular file, but she knew which one it was. Fold. We win.

With the logger's name and the sale data in hand, at least that which was recorded, John and I went back into the night, ready to move ahead with the investigation. One of the next steps was to speak with the owner of the house on the parcel where the stolen logs had crossed, see if there was anyone living in the house, and if there was, if they had seen anything. There was, and he did. The sheriff's office had taken the original complaint and referred it to us since it had to do with timber, a subject of which they, like most police agencies, have little knowledge or experience. But they were willing to help if they could. Since the site was some distance from my station, and since the tenant worked days and was not home until the evening, I asked the sheriff's office to have a deputy swing by some evening and ask the renter a series of questions I provided. They agreed.

A couple of days later, I got a call from a deputy who had the information. "Well, this is interesting. The guy in the house saw some logs come down the hill to the road on a Saturday. And he said they weren't there very long, a few minutes, and that some truck picked them up, but he didn't know who."

Bingo! "That's very helpful," I responded. "Thanks. But what's so interesting?"

"Do you know who lives there?" I didn't. "It's Karl Armstrong." He waited to see if that would register with me. I recognized the name, but I couldn't place it right away. But it hit me as soon as he started to explain.

On August 24, 1970, Karl Armstrong and three others, one of whom was his brother, perpetrated the Sterling Hall bombing on the University of Wisconsin campus in Madison. Though it was done in the middle of the night to avoid injury, a researcher who was still there at almost four in the morning was killed. The bombing was linked to the Vietnam War to protest the school's involvement with the U.S. military, whose Army Math Research Center was located on the lower floors of Sterling Hall. Armstrong was captured years later in Toronto, tried, convicted, and served a seven-year sentence. He had been living quietly since.

My opportunities to meet with and speak with Mr. Armstrong were limited. He appeared at the courthouse, but once the defense attorney learned that we had that piece of information, he stipulated the fact that logs had been removed through the property, such that the witness did not need to testify. While that was a victory for our case, it denied me an opportunity I had been looking forward to. I only had the brief opportunity to thank him for coming in and had to return to court.

We had made the case we needed to. Though the local jury believed the evidence, they decided in the face of Big Brother, the state DNR, to let the loggers off with misdemeanor theft. Not what we wanted, but a conviction, nonetheless.

My brief chance meeting with Karl Armstrong took me back to my youth when the social upheaval resulting from the American involvement in Vietnam tore at the fabric of our society. It had caused many impassioned young people to take dramatic, unwise, albeit well-intentioned actions. Karl Armstrong was one of them. Meeting him caused me to reassess how I had thought, or not thought, about the war. I was a little too young, by just a couple of years, to develop the inflamed response others did. By the time I had grown independent enough to recognize the issues, the war was ending, and Nixon was under fire. Otherwise, I might have felt much of the same outrage that drove Karl Armstrong.

Effective

In another timber theft case, I ran into a defense attorney more accomplished than any I had ever seen in court. He was a very kind and polite man, and he asked the right questions in a respectful way. His strategy was to acknowledge what had happened, which was hard to deny, but to play down the issue of intent and cast a limited perspective on the extent of the damage. He did that very effectively, and after his closing, I almost wanted to stand up and cheer. He was successful to the extent that the loggers were found guilty by the jury but of a lesser misdemeanor charge. In sentencing, the judge only gave them a sentence of probation but required them to pay meaningful restitution to the aggrieved landowner. Both sides won, and we made our point.

The defense attorney was such a decent, effective fellow that I later approached him to help us with an exercise for our forestry law enforcement (LE) annual training. He agreed, which was significant because the training was two and a half hours away and would involve an overnight.

The exercise involved a video of a burning violation investigation, which several of us wrote, filmed, and directed. We shared the video with the hundred or so forestry LE officers at the session. They were to write an incident report, which is the first step toward formal LE action, like a ticket or a criminal charge. This was the last event of the training day, and we spent the evening reviewing the reports, looking for opportunities for a defense attorney to break down their case based on shortcomings or inconsistencies in their reports. The next morning, in front of the group, my attorney friend questioned three of our officers we selected on the basis of their reports. Again, he did so in a very kind, respectful manner, such that the officers watching were not able to generate the dislike that defense attorneys often conjure and were able to learn. It was exceptionally valuable training, and the interview process had its humorous moments. In questioning the last person, another forest ranger, the attorney gradually painted him into a corner on a fact he had observed but had failed to include in his report. When the penultimate question that finally undermined his "case" was asked, leaving him only two directions to go with his response, both bad, he thought for a moment and finally burst out with a loud, inappropriate exclamation. The whole room broke up. It was the perfect ending to the session. The point had been made in a way that no one would forget, and everyone, including the three guinea pigs, was both entertained and instructed.

My attorney friend refused any compensation beyond the hotel room we got him for the previous evening. We remained friends over the years, and I used his firm for some personal, noncriminal services years later.

Necklace

One challenging experience began on a day when I had to wade through a field of tall weeds to get to a woods in which I had to do some work. I was inexperienced in identifying plant species outside of forest settings, so I viewed the growth in the field simply as a small impediment between me and the woods. I was to find out the next day that the impediment had a few ideas of its own. An irritation developed on my upper chest and lower neck around the margins of my T-shirt. The next day, the irritation turned into a burning sensation, and blisters soon followed, some of which were as much as three-quarters of an inch across. Again, someone more experienced quickly identified the blisters as the product of poison parsnip. I did some research so I would better recognize the plant and was committed to not encountering it in that way again. The blisters and the burning sensation did not give way to scars for almost two weeks, and they were still visible a year later.

Poison parsnip, *cicuta maculata*, is in the carrot family and is extremely poisonous. Also known as water hemlock, it contains a neurotoxin concentrated mainly in the roots. Some Native Americans called it suicide root. Considered our most toxic plant, it can cause all sorts of ghastly symptoms, even leading to death. Another name for it is cowbane, as livestock have been known to die in as little as fifteen minutes after having consumed it.

GITTIN' 'ER DONE

One of the other features of field forestry that struck me was the great variety of seed strategies for perpetuation employed by different species of trees. On one end of the spectrum, walnuts produce fruits (the plant part containing seeds) nearly the size of a baseball, albeit in relatively small numbers. These seeds are rugged and durable and, because of their size, can withstand extreme weather and still provide adequate resources for a considerable period for an ambitious shoot to form the beginning of a future tree.

On the other end of the spectrum are the elms. I am old enough to remember, from my youth, streets under canopies of American elms before Dutch elm disease eliminated them from the landscape. Elms produce fruits by the millions. They are tiny, flat, and no more than a quarter of an inch across, most of which is a wing that surrounds the even tinier seed in the middle, enabling it to float on even gentle air currents some distance, a scheme to spread the species to new areas, away from the parent. The vast majority of these elm seeds never even germinate. My dad had me shovel up and bag them by the bucketful. But there are so many of them that even the tiniest percentage of success guarantees the future of the species. This strategy is the opposite approach to that of the walnuts.

In between the elms and the walnuts are fruits of varying sizes and different ways of disseminating themselves. Every kid is familiar with the helicopter-like fruits that the maples generate and the great fun watching them whirl down in the breeze, trying to catch them before they hit the ground. The ashes and the basswoods each employ a different sort of wing to reach out and spread their influence. One plant produces a tasty seed that the birds like to eat but which has a cathartic effect. So, the bird eats the seed, flies away to a new location, and "liberates" the seed. However far away the bird has flown, the seed is ready to claim a new location.

OPEN SESAME!

Serotiny. It is not an unfortunate, obscure condition to be treated. Serotiny refers to a tactic where fruits remain closed and do not release their seeds until a particular time. In Wisconsin, the jack pine (*Pinus banksiana*) employs this strategy. Sealed cones can be seen on young jack pines (because they're low enough to be in view), tightly gnarled around the seed. In a fire, the cone gradually opens, taking its time to allow the fire to pass. And to further ensure the seed's success, there is a small restrainer, a sort of gate, that restrains the seed until a rain or two has caused some expansion and contraction, causing the gate to detach from the wall of the cone and only then allowing the seed to be released. So, the seed hits the ground not when the fire is present but after the fire has passed and a couple of stage-setting rains have occurred. Pretty clever, huh?

As Good As It Gets

One year, part of my forestry assignment was to provide some help to the Richland County forester, who had an inordinately large number of Manage Forest Law applications to survey and process.

Normally, county foresters keep the most interesting applications to do themselves and pass on those of less interest to someone helping, like me. So when I went to the town of Willow to do the fieldwork for a twenty-acre application, I wasn't expecting much. This was an unusually small acreage, and I assumed it was a piece that had been mistreated in the past, and the landowner was now simply looking for the tax relief the MFL program provided in exchange for following a management plan.

I had to walk through a small open area to get to the parcel (avoiding any poison parsnip). But when I began to take in the trees, my jaw dropped. It was some of the finest mature timber I had ever seen. Red oak, white oak, sugar maple, basswood, and an occasional elm, averaging more than thirty inches in diameter, filled the forest. I had seen something like this in Hoosier National Forest in southern Indiana when I was an undergrad, but never in Wisconsin. My sampling process generated estimates of board feet volume more than twice anything I had ever measured in Wisconsin before. A check of the rings revealed that their growth rates had begun to diminish, meaning that this timber was very close to biological maturity. As such, for the twenty-five-year window of the plan, it would have been irresponsible of me not to recommend a harvest. I did so. But the thought of that magnificent woods falling to the saw was almost more than I could contemplate. The bolls of the trees were arrow straight and clear, the crowns spreading out fifty feet above the forest floor. The canopy closure was so complete that there was very little sunlight reaching the forest floor, and ground vegetation was quite sparse. One could have walked through this woods comfortably and safely barefoot. It was a magnificent sight to behold, and I can still see it in my mind.

I was so impressed that I called my colleague in Richland Center when I got back to the office.

"I thought you'd like that one," he smiled through the phone.

PRESCRIBED BURNING AND OTHER FUN

FIRE AND RAIN

They say opposites attract. Many ecosystems depend upon a good mix of opposite conditions; prairies are a good example, needing both hot, torrid fire and cool, refreshing rain. If they do not experience fire, they are soon displaced by other aggressive pioneer plants that are not tolerant of fire. No fire equals no prairie. But by the same token, once fire liberates from those invaders, they need adequate rain for the remaining plants to flourish. Similarly, oak savannas and all forest types with significant components of upland oak require at least occasional fire to maintain their oak mix.

Many tree species grow in a wide variety of circumstances but have no tolerance for the presence of fire at all. Red maple is a good example of a tree that grows under a vast range of conditions. But with its thin bark, it is highly susceptible to even light, cool fires, especially when it is young. Oaks, on the other hand, with their thick, corky bark, are not significantly influenced by fires, especially the relatively cool ones that can be conducted deliberately under prescribed circumstances. But like many trees, once the competition is eliminated, the oaks need rain to thrive.

I always found it interesting to watch news broadcasts after significant forest fires we had worked. Like everyone else, we got a kick out of the chance to see ourselves on TV. What piqued my attention was the way news broadcasters generally characterized the area that has been subjected to fire. Normally, the lead line was something like this: "Fire today destroyed 26 acres of woodlands in such and such a township in Sauk County." And the story would go on from there. With an admittedly condescending attitude, I would say to myself, *No. Those acres weren't destroyed. They're still out there. I can show them to you.* What *was* true was that some of the vegetation on those acres had been destroyed or damaged, and the whole scene was a dark, dirty, discouraging mess, especially with a bunch of dirty and tired firefighters around. To the untrained eye, all appeared to have been lost. But the news folks never came around a few weeks later, or the next spring, or a couple of years later, to see the burst of new life that took advantage of the opportunity created by the fire's effects. That sort of response was beautifully apparent in Yellowstone National Park after the extensive fires that took place there in 1988. For a couple of years, there was nothing but large expanses of tall, dead, lifeless stems. But soon, there bloomed a fresh, new, vibrant undergrowth of trees, elbowing one another to stretch up to the newly available light. And with them came a surge in the numbers of animals that thrived in the close cover those young trees provided. Those scenes burst with life and totally refuted the notion that those acres "had been destroyed." The time after a fire is a good testimony to the determined nature of life.

Firing Black Hawk

For years, the 835-acre property had been privately held. Back in the 1830s, what is now known as Black Hawk Ridge was the site of the Battle of Wisconsin Heights between state militia forces and Sauk and Fox tribes under the leadership of Black Hawk. In its latest commercial manifestation, it provided cross-country skiing and was a place for retreats for various groups. For a while, food was available there.

For whatever reason, the business closed, and the property was made available to the state. It fell into the hands of the DNR and became a unit of their Lower Wisconsin State Riverway. The DNR began a slow, careful, methodical redemption of the property toward something nearer its pre-settlement state. Now heavily wooded, Native American-constructed effigy mounds suggest that parts of the area had once been much more open and that there were areas within the ridge property that had been used for sacred purposes.[14] The north end of the property had clearly once been open ground, perhaps prairie, and now was covered by small trees, dense shrubs, and miscellaneous ground cover, much of it non-native. There was discussion of doing cultural work there, that is, removing individual plants and invasive species by hand. But that would have been so time-consuming that we decided to conduct a burn on the property, just as it was, and see how much benefit could be gained in that far less labor-intensive way. There was little to lose if a controlled fire would help. So, the burn unit boundaries were identified, and containment lines, which were a couple of feet wide and cleared

---

14 An effigy mound is a mound of dirt constructed by Native Americans. They are created for sacred ceremonial purposes and sometimes serve as burial sites, and can be larger than a football field. Often in the shape of an animal or another figure, they represent a connection to the earth and to life. They are protected by law.

down to mineral soil, were constructed all the way around the unit.

The day set for the fire was ideal. It had been quite dry, and almost all the dead fuels in the burn unit, even the larger ones, were dry enough to ignite. That meant we would have a hot fire, which would be effective in eliminating invasive plant species, some of which had already started to leaf out, using up scarce energy reserves in their root systems and making them quite vulnerable. At the same time, the weather on the day of the burn was both warm and dry, but the wind was limited so that once we burned an adequate break all around the unit, we could let it burn slowly, most effectively eliminating much of the dense vegetative cover.

It doesn't happen often, but this burn was perfect. It burned hot and hard, but there were no escapes—nothing even close. Because of some of the larger fuels that got involved, it had to be monitored for a couple of days until things burned out. But that was okay.

What turned out to be most interesting was what researchers found in the following weeks and months. The fire did eliminate much of the built-up fuel on the site, but it was also disclosed that there was previously some quality prairie component there, albeit severely stunted. Those residual plants responded in a way no one could've anticipated, shooting up several feet tall the following spring. But beyond that, an astonishing number of artifacts were revealed. Arrowheads, tools, cooking equipment, and various other items were discovered, but not all of them were Native American in origin. Some items were obviously sourced with American militia forces, indicating the specific site of the bloody battle with Black Hawk. Those discoveries led to the site being named to the National Register of Historic Places. It is the only remaining intact battlefield from the American Indian Wars in the Midwest. All that from a little fire.

Let 'er Rip!

My job allowed me to work alongside The Nature Conservancy, a thoughtful and well-managed nonprofit organization focused on preserving quality examples of increasingly rare biological ecosystems or plant communities.

They determined that there was a very high-quality dry prairie north of Spring Green, immediately north and west of where I lived. I used to take walks there. The light soils, sandy from long past glacial outwash, left the opportunity for upland prairie plants to establish themselves. In the absence of fire, part of the prairie, on a steep south-facing slope, had become invaded with red cedar trees in recent decades. They were actually a species of juniper requiring very little moisture. They had grown so dense that they generated almost complete shade on the ground beneath, severely stifling the native prairie species. Over the course of several winters, The Nature Conservancy organized weekend workdays, mostly staffed by volunteers, during which the cedars were cut, piled, and burned in the snow. Eventually, they cleared the entire hillside, allowing the prairie underneath to begin to recover.

Soon after that, the need to burn the prairie arose, which meant that I got involved, for which I was delighted. It was always a pleasure to work with The Nature Conservancy because they did everything in a highly organized and careful manner, including prescribed burning. I had to be present to issue them the burning permit, and I hung around to monitor their activities, albeit with no operational responsibility, since at any moment I could get called to an out-of-control fire. It was just stimulating to be around any well-managed burn.

The fires on the hillside were especially pleasurable because there was almost no danger of them escaping. Since the slope was south facing, angled toward the sun, it heated quickly in spring, and the snow melted there before it did any place else. To the north of this hill was the backside of the slope, either level or north facing, and quite wooded. As such, snow would persist there for weeks after the prairie was free of snow. Because there were some houses to the south, the burn was always done when there was a south wind, so the embers from the burning prairie would fly over the hill and land harmlessly in the snow. The fuels on the slope were thin enough that the slow creeping burn needed on denser, more matted prairies wasn't essential. So, the prairie burn unit was lit at the bottom, near the base of the hill, and we all got the thrill of watching eight- to twelve-foot-high flames scream up the hill, fearless about the escape potential. There was a little mop-up to be done afterward, so I invariably found a reason to move on soon after the main fire had completed its run up the hill. After all, I didn't *want* to do mop-up if it could be avoided.

How Could This Happen

Early in my career, I had a very humbling experience. Actually, I had quite a few, but this one involved a prescribed burn.

We had lived in the little log house, the home we moved into when I started in Spring Green, for a couple of years. The year before, we had started a garden, which had been very successful, and we decided to expand it. To do so, I needed to clear off some grass on the far side of the present garden, between the driveway and the pine woods, and the easiest way seemed to burn it off. It was only about three hundred square feet, and I think, in retrospect, it would have been easier to mow. But I was a fire guy, and I wanted to burn it. I brought home two back cans, just in case.

I went out after six, like the law says, to make quick work of my little burn, having written myself a permit. The wind was pretty calm, the grass was sparse, and I had difficulty imagining how there could be a problem. With no wind, it didn't seem to matter where I lit the fire, so I lit it by the garden on the driveway side, away from the pines. Ever so slowly, the fire crept through the first grass, accomplishing just what I wanted. A puff of air came along and moved the fire into a patch of fresh grass. It advanced a little more quickly than it had before, but I was okay with that. *Let's get this over with.* But then there was another puff, and this one sustained itself. *Okay. Good. We'll be done soon,* I thought. I positioned myself to squirt out the fire with a back can when it reached the pines, as the hose itself wouldn't quite reach the edge of the burn.

Suddenly, the breeze picked up a little bit, and the fire was moving a little more quickly than made me comfortable. Since it was almost done anyway, I decided to squirt it out with one of the cans. Starting at one end, a couple of good squirts put out a third of the edge. But then something clogged the nozzle. I compressed the pump harder and was able to push something through and got a little water. But it was still jammed. The fire had now moved past the edge of the grass I intended to burn and was now progressing through fresh grass toward the woods, still with only ten-inch flames. My efforts became sharply focused. I quickly removed the nozzle and cleared out whatever impediment was there. Quickly replacing it, I anxiously suppressed more flames, which had expanded. But there must've been dirty water in the can because it jammed up again almost immediately. I set it down roughly and ran for the other can but experienced the same problem. By then, the fire had reached the edge of the woods, and what had been mundane had quickly evolved into terror, complete with paranoid speculation about how I would live down having to call the fire department to the forest ranger's house to put out a fire in the pines that he, himself, had started. Would I lose my job? Would I have to move away?

As quickly as it had come up, the breeze quit, the flames dropped, and I was able to stamp everything out with my boots. The whole episode couldn't have taken more than four or five minutes, but it gave me the opportunity to understand what people I encountered on escaped fires must have felt like as they witnessed the development of their catastrophe. I grew a substantial empathy appendage that day.

What Does *Your* Daddy Do?

The benefit of living in a small town and being known sometimes shows itself in unusual ways.

When my older son Ben was seven or eight, I had the opportunity to help (off the clock) with a nearby prescribed burn that took place in the evening. For fun, our whole family came along. I was tasked with starting ignition on one of the flanks after the main backing fire had been ignited and burned for a while. Jan had little Pete in her arms, and Ben walked along with me as I lit the side of the fire with a drip torch.[15] The burn went as planned, and a good time was had by all.

A week or so later, I received a phone call from Ben's teacher, who was also a friend from church. Through chuckles, she let me know that there had been a session in the classroom where the kids were able to talk about their parents' work. When it came around to Ben, he proudly announced, "My daddy starts fires."

In some settings, that might prompt a call to the local police or Family Services agency. Not in Spring Green, my community. All that was elicited there were gales of laughter.

When the Levee Breaks

"If it keeps on raining, the levee's going to break.
When the levee breaks, have no place to stay."
—Kansas Joe McCoy and Memphis Minnie, "When the
Levee Breaks," 1927, covered by Led Zeppelin, 1971

The peculiar specifics of history sometimes lead to unusual practices and arrangements. The management of the levee at Portage was one such example.

Lying along the Wisconsin River, Portage is an old city (by North American Midwest standards) at a low point where the Fox and Wisconsin Rivers are only a couple of miles apart. Historically, this made for a strategic trading opportunity between the two commercial arteries, the Fox, eventually connecting to the North Atlantic, and the Wisconsin, to the Gulf of Mexico. To better facilitate the flow of goods between them, a canal was built across the portage (Portage, get it?). The canal is no longer in use, and there is a large sand "plug" at the Wisconsin River end, preventing it from flowing into the canal and Portage's "First District." Today, much of the city occupies the higher ground to

---

15 A drip torch is a tool used to start a line of fire, as when burning a prairie. It is a columnar tank with a handle, holding about a gallon. From the top extends a twelve-inch tube with a wick on the end. In use, the wick is lit, and the can is held upside down, the fluid drips the wick, is ignited, and "drips" onto the fuel, setting it afire. The rate at which fluid is released is controlled by an adjustable valve that allows more (or less) air into the tank. The fluid mix is generally about one-third gasoline (for the punch) and two-thirds diesel (for staying power). One of the first times I used a drip torch, my boss instructed me to sit on the tailgate of the pickup while he drove along the edge of the parcel to be burned, with the valve wide open. And he was an aggressive driver.

the north. However, the so-called First District, in the southeast corner of the city, is both populated and at risk for potential flooding, especially from the adjacent Wisconsin River via the abandoned canal.

At some point a levee was built on both sides of the Wisconsin, west of Portage, to protect developed areas from flooding. On the east, or city side, it cuts a somewhat festive profile. Accompanied by signs and topped with a paved hiking/biking trail, it lies just across Highway 16 from bustling businesses: car dealers, restaurants, gas stations, and even a lumberyard. On the west side of the river, the lowlands are more pervasive, and the levee system is much more extensive. Its path winds more quietly through lowland forests with only scattered collections of homes. There is a drivable path on top, too, but it is dirt and feels more remote, more isolated, quieter, in an almost primordial forested setting.

As water levels in the river rise, various protective protocols progressively come into place. When the flooding threat reaches a certain level, the DNR is brought in to monitor and address potential problems as they develop. Why the DNR? Good question. I never got a clear answer, but it was a practice set in stone. The forest fire folks were well equipped for such a task, with our organization skills developed for fires. So when flooding risk materializes, the question is never, "Why?" or "Who?" but "How?"

The Wisconsin River's flow is interrupted by a long series of dams and associated reservoirs. While the dams' principal purpose is to generate hydroelectric power, they can also serve to help control flooding when there is rapid snowmelt or heavy rainfall. When water levels get high in one location, they open the dam to release water downriver, where levels are lower. It takes several days for high water in the upper river in northern Wisconsin to make its influence felt in southern Wisconsin. So, dam alleviation activities happen in highly considered and coordinated fashion. But when it rains everywhere, all at once, there's no place to put the accumulating water, and agonizing decisions about where flooding will have the least impact have to be made. Being near the end of the series of dams, the effects of statewide rains on water levels linger longer in Portage.

The rains that dance in and out of spring bring an end to the forest fire season. But one year, they didn't just end it; they were so intense that they washed it away, and they did so statewide, all at the same time.

The river had been quite low, and when water levels began to rise, the hope was that the necessary monitoring could be managed by the forest ranger in Poynette and a few other local DNR employees. Soon, however, intuitive concerns about flooding were confirmed by reports on water levels and dam settings upriver. Water was rising everywhere, and there was no alternative to releasing significant amounts of water downriver. So we geared up to respond fully, which could eventually involve around-the-clock levee monitoring by multiple teams.

Under normal circumstances, the management of the levee involves occasional visual monitoring and "control" of burrowing varmints, like groundhogs and the like. A holey levee is a compromised levee, so any holes identified are filled with dirt. As things worsen, and seeping is identified on the outside of the levee, a fabric specifically designed to retain the solids (soil) and release the seeping water is tacked over those places. Seeps can lead to trickles, which can lead to holes, which lead to flows, which lead to catastrophic failure. And with the high pressure of deep, fast-moving waters, that process can happen with disturbing and dangerous swiftness.

On this occasion, we worked more and more closely with the city. They were responsible for keeping understandably anxious folks in town informed. At max levels, the DNR provided a public information officer to report to the governor's office. But the city was responsible for making evacuation decisions and enforcing them. Our job was to monitor the associated safety concerns and water levels on the ground and provide intelligence. Water levels rose even more rapidly than anticipated. That was alarming, but it also meant that we would morph into nighttime operations immediately. We otherwise try to avoid night work on the levee, but the need now outweighed safety concerns.

Monitoring crews soon shifted from one person to two. Eight-hour shifts involved three teams instead of one, and their activities had to be closely monitored. That is a more active process than it sounds. A team's problems can be indicated not just by communicated concerns but also by no communication at all. So someone had to be at the other end of the radio, responding to needs but also assuring that not more than fifteen minutes went by without each team checking in. Teams were armed with rifles; at that stage, any burrowing animal spotted on or near the levee was shot on sight.

Supplies, including water, food, fuel, batteries for lights, and patching fabric, had to be located and made available to those teams at a place convenient to them. Under such emergency conditions, local businesses often accommodated those needs *gratis*, and they made themselves available during off hours. It was another example of an effective community; everyone was in it together.

Because circumstances changed so quickly, I had to oversee activities through the night, too. One of our newer rangers, a very capable young man, had been given the afternoon off with the understanding that he'd be in Portage through the night, overseeing field operations.

"Am I ever glad to see you," I greeted him. It was almost seven. "Are you ready for this?"

"I think so. I've never done anything like this before."

I tried to be reassuring. "I never have either. But it's a lot like overseeing crews on a night fire. You'll get comfortable with it quickly."

"Okay." The ranger still looked unsure.

"I'll be around if something comes up. I'm going to get a room and get a little sleep." Time to get to specifics. "Here's an access card for the back door to the city office building. That's our headquarters, and they lock the main door after hours. Ralph is across the river, so we'll go find him, and he can brief you before he takes off." Ralph had done in the afternoon what Aaron was going to do through the night.

This was starting to sound more familiar, like protocol on a fire, and I could see his tension begin to ease.

I continued. "Then we'll go and connect with each of the crews you'll be overseeing during the night."

"That'll be good." He was ready.

We jumped into my truck and drove over the bridge across the Wisconsin to find Ralph. "Wow!" He got his first sense of what we were up against. "No sandbars." And then, "I can't even tell where the edge of the river is supposed to be."

"And it's still coming up. After we go back to the office, I'll show you how to track flows and levels upstream. That gives you some idea how fast it's going to rise and maybe for how long."

We met Ralph, who took Aaron out on a nearby section of the levee and showed him the trail and an area that had been patched. While they were there, one of the monitoring teams roared up on a four-wheeler, reporting all to be well, both to Ralph and to Aaron, in his new role.

When we finished, I released Ralph to head home, and Aaron and I continued.

"One crew down. Two to go."

"Are those guys local?" he asked.

"Yeah. They're from the game farm. They've done this before. One of the other teams is, too, and the other one is wildlife people from Dane County, I think."

"Not much room for error with those four-wheelers on the levee."

"You're gettin' it," I confirmed. "That's why we don't really want folks out there in the dark. But when it gets like this, we don't have much choice."

We found the other crews and headed back into town. I showed him the flow data source on the computer there and where all the phone numbers were.

"These crews started at four, so crew change is at midnight. The old crews won't leave until the new crews are there and briefed. When that's complete, I want you to call me. I'll have my phone by the bed and on."

"Don't you want to sleep?" he asked, considerately.

"I'll sleep better if I know things are going smoothly. So call me, too, if there's any sort of issue. If you're not sure if you should call, call. Okay?"

"If that's what you want. Will do."

"Thank you. You'll be here until seven or eight or whenever

Ralph comes back. I'll be up and here before then. Questions?"

"So, if something happens, I should call you, right? Should I call someone else, too?"

"Good question. Yes, call me. And there should be someone from the city around the office, at least off and on, through the night. Keep them apprised of anything that happens. Or if everything is okay, tell 'em that, too." Right on cue, someone from the city manager's office walked in. Introductions were made, and Aaron seemed pretty comfortable.

"Any other questions?" I offered a last chance.

"I can't think of anything." He smiled. "But I know who to call!"

"There you go. I'll call you before I crash in case you think of anything."

"I'll be okay." He may have sensed I doubted him.

"I know. I know. There's just so much at stake. And I'm new to this, too." In a pause, each of our minds drifted to visions of disaster scenarios if a levee should fail. Commitment reinforced; we parted without further discussion.

Before I turned in, I went for a walk on a section of levee on the west side of the river, away from town. There was still some light, and I wanted to take a last look at what we were dealing with at the more recent, higher water levels. I chose a section that wended its way through the eerie lowland hardwoods, a somewhat spooky environment under "normal" conditions. Always humid, lowland forests lie beneath most breezes, so they tend to be quiet and still, except for the birds. The prehistoric cry of the pileated woodpecker conveyed a *Jurassic Park* ambiance. Mix in the musky smells of downed wood that decays so quickly in this environment, and the sense is unlike anywhere else in Wisconsin.

The water was still three feet from the top of the levee, still a "comfortable" level, but at least a foot higher than my last visit, with worse conditions anticipated. And it was moving much more quickly than it had been, swirling, undulating relentlessly, like an exotic eastern dance, all in almost complete silence. It was disconcerting to imagine what would happen to anything, anyone (like our patrol crews) who fell prey to those dark, churning waters. They would be drawn under and out of sight almost immediately, only to be found days later a hundred yards or a hundred miles downriver. The safety concerns were real.

I had been scheduled to manage things the next day, but that was before we knew I would have to be there through the night. There is always a certain amount of confusion with these sorts of operations, especially when they are transitioning to full level. The owner of a motel gave us a carte blanche, as-needed option on rooms. Day and night morph together, work shifts stretch out, and you snooze when you can.

Over the next couple of days, water rose to precarious levels. We were in constant communication with engineers upriver, monitoring water levels and rain chances there. We were also in direct contact with local weather forecasters about rain chances in the Portage area. The river was within a foot of the top of the levee on the west side. Homes were more sparse there, but there were quite a few of them. The city had no authority in the township; it would be the county sheriff who would need to make evacuation decisions. So they were part of all the discussion, too.

Water that would have otherwise drained "down" into the river was now starting to "pile up" in these neighborhoods. Roads were more than wet, and there were pools of water in some areas. Surely, some homes had water seeping into basements. Residents were concerned; their memories and history were all contained in their homes. Times like these change perspectives.

By the next day, another potential problem came to light. The "plug" by the canal, which seemed so imperiously powerful a few days before, looked significantly more vulnerable. We had walked on it previously, and it was as sound and solid as Gibraltar. Now, though, the water was near its top, and a toe touch yielded a consistency more akin to dense Jell-O than any rock. It was so tenuous that a trough cut with a butter knife could trigger a breakdown with calamitous consequences. It would flood the canal and more, inundating the first ward in water. The city posted an armed guard to prevent disturbance. And a recommendation (not an order) was made to all residents living in the first ward to evacuate. That meant increased law enforcement patrol there to preclude any possible looters. The issues stack up quickly in ways one might not anticipate under these circumstances.

Amid tiptoeing, held breaths, vigilance, and prayer, we got through the next few days, and water levels began to recede. One of the encouraging things about such potential disasters is that they go away at least as quickly as they come. I'm sure that our efforts in patching and sustaining the levee prevented serious flooding.

The operation continued for a while. Supplies needed to be restocked. Areas of weakness on the levee had to be addressed more permanently, and people, workers and residents alike, needed to get some sound sleep. Weaknesses in the operation were identified, and solutions were instituted for the next time, and there's always a next time. Reports were written, attaboys and attagirls shared, and there was a general sense of thankfulness. Catastrophe was avoided, and some new relationships were established, even if only for a short time.

Those connections don't fade easily. When so much is at stake, experiential memories are etched more deeply into the folds of the brain. Several years later, I ran into a Portage city official whom I hadn't seen since the near-flood. We exchanged little more than a "How are you?" but our link had been deeply forged in the crisis, in uncertainty, with the need to develop trust in a hurry, and our smiles stemmed from the recollection of how effectively we worked together to do genuine good for people under stress. Those are satisfying moments. I still feel a smile start to creep across my face when I think back.

EPILOGUE

We first run into trees in the creation story in the first chapter of Genesis. Trees are special, and they always have been. Whether it's for the initials marking a long-lost love affair, the base for a rope swing into the river, or a memorial to some person or event, trees have long been our companions, our cronies, our amigos through important parts of our lives. There is richness and blessing that flows from being able to spend work time among them, whether it's trying to create better growing conditions for them or protecting them from the ravages of fire. And when you spend enough time with them, you learn that, like people, each one is different, distinct in some way. For me, they were an encouraging constant in a good work life.

Soon after I retired in 2011, many aspects of the DNR's fire program changed. Some of the changes made some limited sense from a cost perspective, but they deeply influenced the culture of the fire program in Wisconsin. And that culture was a large part of what drove its success.

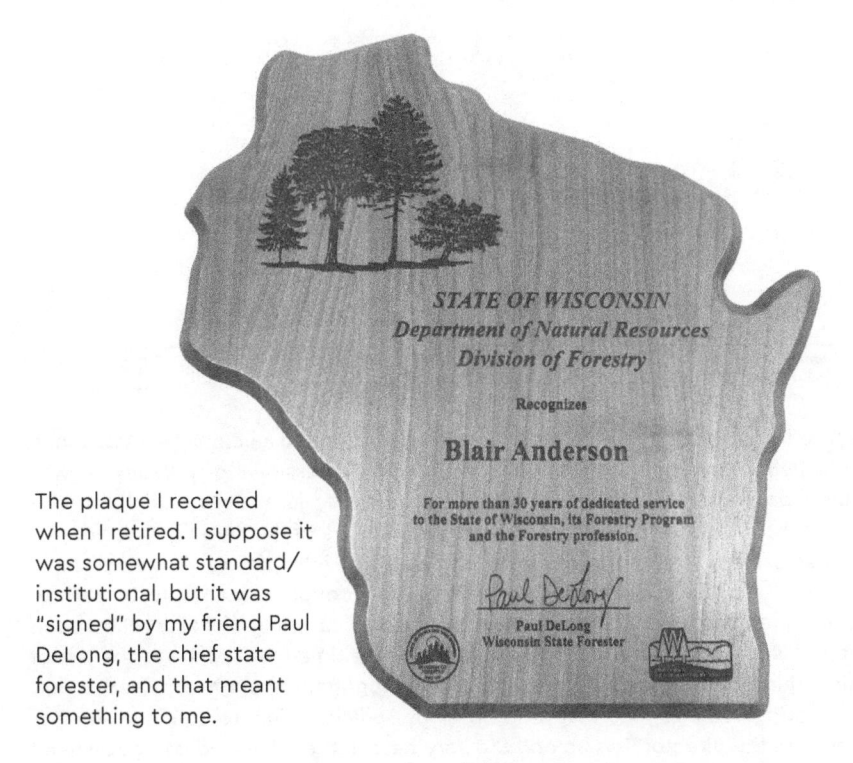

The plaque I received when I retired. I suppose it was somewhat standard/institutional, but it was "signed" by my friend Paul DeLong, the chief state forester, and that meant something to me.

BLAIR W. ANDERSON

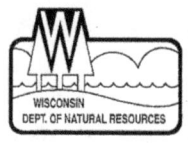

State of Wisconsin \ DEPARTMENT OF NATURAL RESOURCES

Scott Walker, Governor
Cathy Stepp, Secretary

101 S. Webster St.
Box 7921
Madison, Wisconsin 53707-7921
Telephone 608-266-2621
FAX 608-267-3579
TTY 608-267-6897

March 17, 2011

Dear Blair,

As you retire I want to express my deep appreciation for your many years of service to your fellow DNR employees as a volunteer Employee Assistance Coordinator (EAC) for the Department.

You have been an EAC since the beginning of the Department's Employee Assistance Program in 1983. In fact, you are the last of the original group of "LRC's" – Local Resource Coordinators, their name in the 1980's. Over the years you have assisted co-workers to more effectively cope with a wide array of personal problems and crises. You have been there to listen, support, problem solve and when needed, refer people on to further help.

Yours was an unsung role and quiet role. At times you were privy to very personal details of coworkers' lives. Your confidential sensitivity to their vulnerable emotional state was comforting to them. No doubt you saw some tears shed by people you knew and respected. That is not easy.

Your reward for this service was intrinsic, the knowledge that you helped fellow DNR staff through some of the toughest times of their lives. Their "thanks" and a smile, or a silent nod in the hall or elevator spoke volumes of gratitude to you. You performed this service even if you were having a bad day, or were trying to meet your own deadlines. You took the time to assist your coworkers

I will miss your low key approach, your sense of humor, the twinkle in your eye as you grin and laugh, and your insightful comments. Have a GREAT time in your retirement. Stop in and say hi if you ever venture my way. Take care.

Sincerely,

Jeff Carroll
DNR Employee Assistance Program Director

Adios Blair! I will miss you.

www.dnr.state.wi.us
www.wisconsin.gov

*Quality Natural Resources Management
Through Excellent Customer Service*

A few years after I started work with the DNR, they implemented an Employee Assistance Program to help employees with issues in their lives, professional or personal. It was staffed by a paid director and volunteer participation by interested field folks who served as field contacts. I was nominated and became one of the field contact people. While contacts often resulted in referrals, just as often, folks simply wanted to have someone to talk to about whatever they were going through, someone to empathize with them. Sometimes we prayed together. I usually only had a handful of contacts each year, but it was a nice opportunity to be helpful. By the time I retired, I was the last of the original thirty-two contact people still in the program. Others had retired or left the agency or moved into jobs that didn't provide sufficient time to continue serving. This letter from the director was among the very best of the "thank-yous" I got when I retired.

Retirement Resolution

Blair Anderson is retiring from the Department of Natural Resources after 31 years of dedicated service to the State of Wisconsin.

Blair began his career with the Department of Natural Resources in July of 1980 as a Forest Ranger, serving 21 years in that position in Spring Green. In 2002 he promoted to a Team Leader for the Dodgeville area within the South Central Region. He served diligently in that position for 6 months before being asked to leave. So he became the Section Chief of the Fire Management Section within the Bureau of Forest Protection in 2003. During his lengthy career, Blair dedicated his efforts towards protecting the natural resources and safety of the residents of, and visitors to, Wisconsin. This dedication was evident when he single handily assisted a number of bikini clad women suppress a fire that had escaped their control, going so far as to refuse aid offered by others, who were wondering why he was tied up on the fire for so long. On April 1, 1999, while on stand-by, Blair was diligently monitoring the brush pile he was burning in his backyard to ensure that it did not escape and become a forest fire, to the point of not hearing his phone ring over and over again. Because of his diligence, his brush pile did not burn out of control. However, at the same time, and the reason his phone was ringing, the Avoca marsh was set on fire by an arsonist and burned 1,350 acres.

Throughout his career Blair worked on building relationships outside of the Department including serving on the Great Lakes Forest Fire Compact and Wisconsin Interagency Fire Council. He also built strong relationships with local law enforcement agencies that paid dividends when he investigated a large amount of smoke coming from the Town of Spring Green waste transfer site. Disregarding his own personal safety, Blair walked through the heavy smoke to the origin of the fire, which was the Sauk County Sheriff's department burning confiscated marijuana. Because of the relationship Blair nurtured with the Sheriff's office leading up to this incident, they willingly gave him a ride home, via the Culver's drive thru to curb Blair's sudden appetite, so that he did not have to drive his state issued fire truck under the influence of a controlled substance.

WHEREAS, Blair Anderson has served the people of the State of Wisconsin and well in the performance of duties and has thus earned the commendation due such an exemplary public employee;

NOW, THEREFORE, the Forest Fire and Law Enforcement Section recognizes such faithful service with due respect and gratitude, and wishes Blair a healthy, happy and rewarding retirement. The Forest Fire and Law Enforcement Section also offers their deepest sympathy to Blair's wife, Jan, in having to put up with him around the house more.

This tongue-in-cheek resolution was created by some of the delightful folks with whom I worked throughout my career. A couple of incidents from which stories in the book sprung are mentioned here.

Other changes, like removing law enforcement credentials from forest rangers, significantly disabled the program's effectiveness. As I watched from retirement, it didn't feel good to see so much of what we had worked for discarded. Some personal losses I experienced caused me to turn my mindset away from the program more abruptly than I would have liked to, and these changes in the fire program enhanced my sense of separation, of loss. I had hoped to do some support work within the program after I retired, but that never materialized through no one's fault. Memories quickly became the principal and most satisfying source of my thoughts about work. That happens to many, if not most folks, but it was accelerated in my case. Switching to "memory mode" was not an option for the younger people who had been on my staff and who continued to work within the program. I felt bad for them, but there was nothing I could do.

Books about experiences or about history are usually more attractive to older readers. Nostalgia develops a stronger pull with time, and older folks may long for fond memories of specific past chapters of life in the past that a story may stimulate. Most of the people in these stories are now retired, if not gone altogether. And those who aren't are not yet of the age when memories become a prime source of reflection. Like most jobs and most activities, the elements that persist most prominently in one's thinking are about people, and people don't really change that much over generations, even if they like to think they do. The settings and the circumstances may change, but the nature of conflict, as well as the opportunities to show grace, are of the same complexion, time over time.

With age, we grow more comfortable with the past, perhaps because we have more of it. We tend to distill out the best and sift out the lesser. Perhaps there is some willing but harmless self-deception at work here. Or perhaps it's our nature; the older I get, the better I am. It's important for us to see the good when we look back on our lives, even if there was less of it than we may have wished. In time, our capacity for change becomes as inflexible as our joints, and the essential importance of brutal honesty in self-assessment, legitimate in our earlier years, begins to diminish. As we grow older, it takes less and less stimulus to steer our thinking to the past. Growing old is not for the faint of heart, and positive memories serve as a respite from encroaching harsh realities.

I somehow stumbled into a job that was good for me, just right. I accomplished some good things, and I was able to identify and develop some personal giftings I hadn't previously been aware of. My sifting of the past has gone well and has yielded some satisfying stories and minimal embarrassment or desire that things had been different. I've had conversations with people who are looking forward to their retirement, double-digit years hence, and am thankful that I never had such a sense that I was hanging on until the end. I enjoyed the work, yet I was surprised at how little I was at a loss the first spring after I retired. Life turned upside down after that, but I managed to get through it. My journey of blessing has continued.

I hope you enjoyed these stories. More importantly, I hope they cause you to reflect more thoughtfully when you see a tree, spend time in the woods, or enjoy a fire. Not only are those activities pleasant in themselves, but they often afford the opportunity to appreciate the distinctions in other people and their perspectives and how much that uniqueness can add to our own experience and perspective.

ACKNOWLEDGMENTS

The creation of *When the Smoke Clears* is the culmination of numerous components coming together.

For this storyteller, one of those components is surely the encouragement of the people who heard my stories and responded, "That's interesting. You should write that down." So, thanks to all those people who endured me going on with the telling longer than I probably should have.

As strange as it might sound, I need to acknowledge the role of the so-called COVID crisis, too. That experience brought about not only an increased availability of hours but also the chance, at my age, to recognize that our earthly lives are finite and that all the "someday" things that I have thought about doing need to get done. And so I began to do what all those faithful friends encouraged me to do: I started writing down the stories.

Speaking with old work cronies brought back to memory some of the old experiences. More stories.

I assembled each story rather quickly because that's the way stories rush back to mind. But later, as I read them, it became apparent that the stories were clearer in my mind than they were on paper (remember paper?). So, I reworked them and made them better. But I realized that I was trying to see what the house looked like from within—a difficult task.

Enter my lovely, ex-English teacher, part-time editor wife, Ann. In addition to her endless enthusiasm and encouragement, she provided significant constructive suggestions on the stories. And early on, as they were still being created, she also provided a good, broad perspective that helped me improve my writing. "Use good verbs" was one that quickly moved to the front of my mind. And so, while you may not know it, her influence has made this much more readable.

The last piece of the puzzle was the title. Thanks to Paul DeLong for suggesting *When the Smoke Clears*. It fit perfectly.

Finally, thanks to my new friends at Little Creek Press, who not only provided the infrastructure to get this from "a bunch of stories" to a book but also provided much-needed encouragement. Shannon Booth, who goes by the humble title of "proofreader," not only addressed my tendency to overuse semicolons but brought fresh eyes to the manuscript. She called me out when I wasn't clear and made smooth the cumbersome in many instances. Again, you, the reader, are the beneficiaries. To her, a "high" five. And Kristin Mitchell brought color and life to a black-and-white manuscript through her inspired ideas for the cover and layout. And if I even make a nickel on this, it will be because of Kristin Gilpatrick's finesse in getting the book into the right hands and presenting it in the right light. Thank you all.